About the Eagan Press Handbook Series

The Eagan Press Handbook series was developed for food industry practitioners. It offers a practical approach to understanding the basics of food ingredients, applications, and processes—whether the reader is a research chemist wanting practical information compiled in a single source or a purchasing agent trying to understand product specifications. The handbook series is designed to reach a broad readership; the books are not limited to a single product category but rather serve professionals in all segments of the food processing industry and their allied suppliers.

In developing this series, Eagan Press recognized the need to fill the gap between the highly fragmented, theoretical, and often not readily available information in the scientific literature and the product-specific information available from suppliers. It enlisted experts in specific areas to contribute their expertise to the development and fruition of this series.

The content of the books has been prepared in a rigorous manner, including substantial peer review and editing, and is presented in a user friendly format with definitions of terms, examples, illustrations, and trouble-shooting tips. The result is a set of practical guides containing information useful to those involved in product development, production, testing, ingredient purchasing, engineering, and marketing aspects of the food industry.

Acknowledgment of Sponsors for *Sweeteners: Nutritive*

Eagan Press would like to thank the following companies for their financial support of this handbook:

National Honey Board
Longmont, CO
800/356-5940

A. E. Staley Manufacturing Company
Decatur, IL
800/526-5728

Eagan Press has designed this handbook series as practical guides serving the interests of the food industry as a whole rather than the individual interests of any single company. Nonetheless, corporate sponsorship has allowed these books to be more affordable for a wide audience.

Acknowledgments

Eagan Press thanks the following individuals for their contributions to the preparation of this book:

Ronald E. Hebeda, Corn Products International, Summit-Argo, IL

Sakharam K. Patil, Cerestar USA, Hammond, IN

Patricia A. Richmond, A. E. Staley Manufacturing Company, Decatur, IL

Karen Schmidt, Kansas State University, Manhattan, KS

Jeffrey Stamp, The Navigator Group Inc., Cincinnati, OH

Laszlo Toth, Western Sugar Company, Denver, CO

Contents

Sweeteners: Nutritive

Chemistry of Carbohydrate-Based Sweeteners

Carbohydrates, or "hydrates of carbon," are an important group of naturally occurring organic compounds. The term describes those molecules composed of the general formula $C_x(H_2O)_y$. Saccharides, the simplest forms of carbohydrates, consist of single sugar units with five or six carbon atoms in ring form. They are commonly called "sugars" or "sweeteners" because they taste sweet. *Monosaccharides* consist of one saccharide unit; *disaccharides*, two units; *trisaccharides*, three units; and *polysaccharides*, many units. *Oligosaccharides* are saccharides with more than three but less than eight units.

The mono-, di-, and trisaccharides are water soluble and have varying degrees of sweetness. Glucose, fructose, and galactose are monosaccharides. Familiar disaccharides include sucrose (made up of glucose and fructose), lactose (glucose and galactose), and maltose (two glucose units). These are described in more detail in Appendix A.

Common polysaccharides found in nature include cellulose, starch, glycogen, hemicellulose, agar, pectin, and lignin. Under ordinary conditions, they are tasteless and insoluble in water. Through enzymatic or chemical hydrolysis, polysaccharides such as starch can be broken down into their component parts (Fig. 1-1). Today, these practices are utilized in the manufacturing of sugars and syrups, which are covered later.

In This Chapter:

Terminology

Reactions of Carbohydrate-Based Sweeteners
  Hydrolysis and Inversion
  Isomerization
  Reduction
  Oxidation
  Thermal Degradation

Monosaccharide—A carbohydrate containing a single sugar unit, usually composed of five or six carbon atoms, existing in a furanose (five-membered ring) or pyranose (six-membered ring) form.

Disaccharide—A carbohydrate containing two sugar units, each composed of five or six carbon atoms in a furanose or pyranose ring.

Trisaccharide—A carbohydrate containing three sugar units.

Polysaccharide—A carbohydrate containing several hundred, thousand, or hundred thousand sugar units (from the Greek *poly*, meaning "many").

Oligosaccharide—A carbohydrate containing four to seven sugar units (from the Greek *oligos*, meaning "a few").

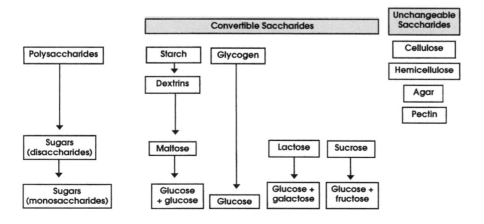

Fig. 1-1. Breakdown of polysaccharides to their component sugars.

Aldehyde—An organic compound containing a -CHO group.

Aldose—A sugar molecule containing an aldehyde group at the terminal carbon position.

Ketone—An organic compound containing a -CO- group.

Ketose—A sugar molecule containing the ketone group at the carbon molecule adjacent to the terminal carbon.

Stereochemistry—The relationship of atoms in three-dimensional space.

Furanose—A five-carbon sugar molecule in the form of a ring.

Pyranose—A six-carbon sugar molecule in the form of a ring.

Terminology

Knowledge of the terminology of carbohydrates is essential in understanding the chemistry, properties, and functionalities of sweeteners. This nomenclature also helps to define the structures of sugars.

The suffix "-ose" in the terms "glucose," "fructose," and "sucrose," for example, denotes that these molecules are sugars. All basic carbohydrate names contain this suffix. Word elements used with -ose have different functions. One is to denote the number of carbon atoms present in the sugar molecule. Trioses, tetroses, pentoses, and hexoses contain three, four, five, and six carbon atoms, respectively. A common pentose is xylose, and common hexoses include glucose, mannose, and fructose. Another purpose of word elements is to denote whether a carbon atom is present in the *aldehyde* form (-CHO) (in which case, the sugar is an *aldose*) or the *ketone* form (-CO-) (when it is a *ketose*). Glucose is an example of an aldose; fructose is a ketose. To completely describe sugar molecules, the two elements are often combined and then the -ose suffix is added. For example, fructose is a ketohexose and glucose is an aldohexose (Fig. 1-2).

Figure 1-2 shows the two sugars in a two-dimensional view called a Fischer projection. It is important to note that these molecules exist in three-dimensional space, and Fischer projections do not tell us anything about the conformation of the molecules in three dimensions. For that, we must look at the *stereochemistry* of the molecules. Stereoisomers are compounds that have the same molecular formula but different spatial organization of the atoms. In aqueous solution, stereoisomers have the ability to bend a plane of polarized light. If the light is bent to the right, the isomer is termed "dextrorotary" (D). If the light is bent to the left, it is termed "levorotary" (L). All sugars containing more than one saccharide unit are termed "D-sugars" when the hydroxyl group nearest the terminal CH_2OH is directed to the right. Hence, the sugars shown in Figure 1-2 are D-fructose and D-glucose.

When a carbohydrate is composed of five or six carbon atoms, it exists in nature as a cyclic form. The aldehyde or ketone group interacts with one of the hydroxyl groups on the carbon chain, resulting in a cyclic acetal (hemiacetal) or a cyclic ketal (hemiketal), respectively. Carbohydrates with five carbons in a ring form are called "*furanoses*," and those with a six-carbon ring are "*pyranoses*." The carbon atom that was previously in the ketone or aldehyde form but has interacted to form the cyclic structure is now a stereocenter. That is, its bonding groups can have different spatial

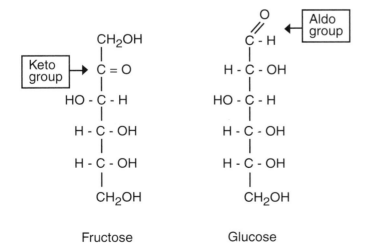

Fig. 1-2. Structures of fructose (a ketohexose) and glucose (an aldohexose).

orientations. In this case, the hydroxyl groups can have two different spatial orientations, or "anomers," termed α and β. Figure 1-3 shows the cyclic forms for glucose in the pyranose configuration. The structures are called Haworth projections when shown in this cyclic manner.

As described earlier, disaccharides are sugars composed of two saccharide units linked together. Disaccharides are typically depicted using the Haworth projection; for instance, the disaccharides maltose, lactose, and sucrose are shown in Haworth projection in Figure 1-4. In this figure, the linkage between the two saccharide units is between the first carbon atom of one saccharide and the fourth carbon atom of the second saccharide, which is called a "glycosidic linkage." The linkage is termed an α-1-4 or a β-1-4 linkage depending on whether the first carbon on the sugar is in the α or the β position. Carbohydrates composed of more than two sugar units (trisaccharides, tetrasaccharides, oligosaccharides, etc.) are linked together in a similar manner. Linkages can also occur between carbon atoms other than the first and fourth. For instance, both 1-4 and 1-6 linkages occur in starch, and 1-4 and 1-3 linkages are found in hemicelluloses.

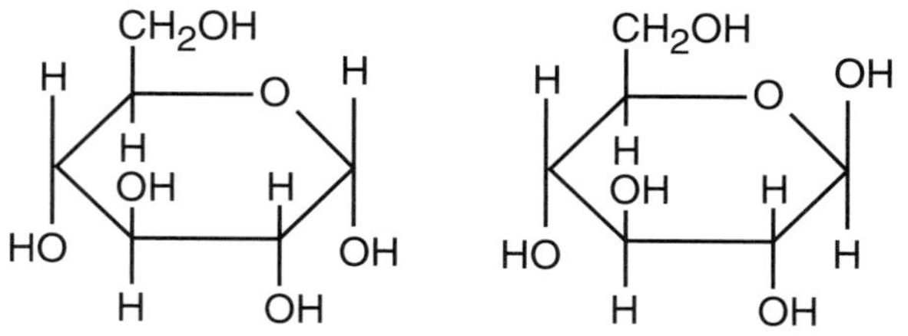

Reactions of Carbohydrate-Based Sweeteners

It is helpful to understand a few of the chemical reactions that are important in carbohydrate chemistry in order to better understand carbohydrate-based sweeteners, how they are made, and how they function in food systems.

HYDROLYSIS AND INVERSION

Our own digestive systems hydrolyze disaccharides, such as sucrose and maltose, to produce simple sugars. Monosaccharides, such as glucose and fructose, are then metabolized and converted into energy.

The hydrolysis of sucrose to its component sugars of glucose and fructose is often called inversion. These hydrolysis reactions are shown in Figure 1-5. The enzyme invertase, often used to hydrolyze sucrose for industrial production, is

Fig. 1-3. Haworth projections of D-glucose in the pyranose configuration.

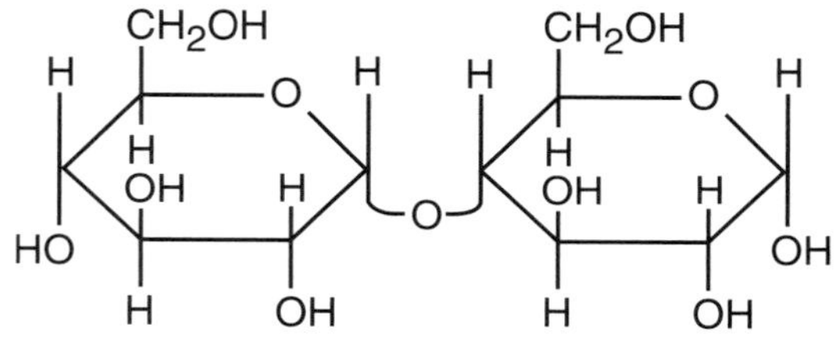

D-Maltose

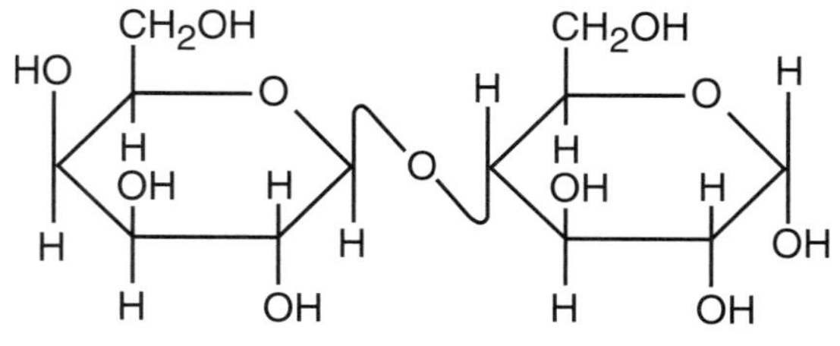

D-Lactose

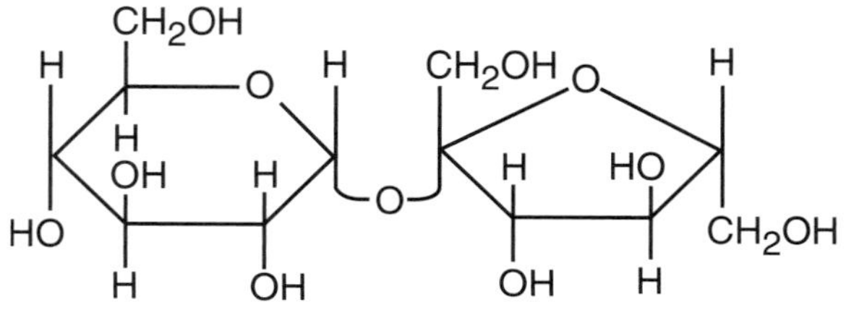

D-Sucrose

Fig. 1-4. Structures of the disaccharides maltose, lactose, and sucrose.

$$C_6H_{11}O_5 - O - C_6H_{11}O_5 \quad + \quad H_2O \quad = \quad C_6H_{12}O_6 \quad + \quad C_6H_{12}O_6 \quad + \quad heat$$

Sucrose: Mol. Wt. = 342.3   Water = 18.02   Glucose - 180.16   Fructose = 180.16

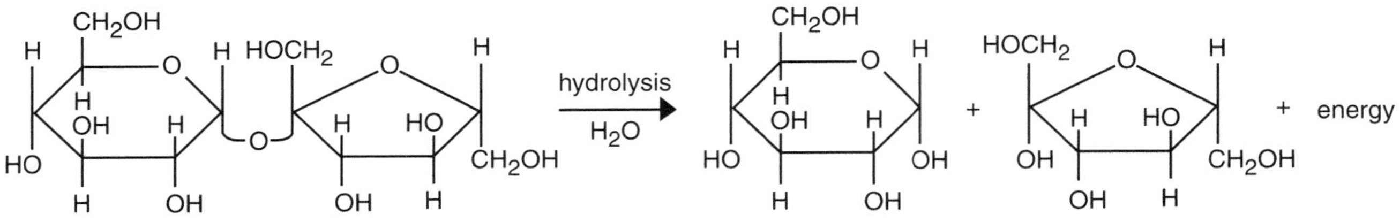

Fig. 1-5. The Haworth structural diagram of the hydrolysis of sucrose into glucose and fructose.

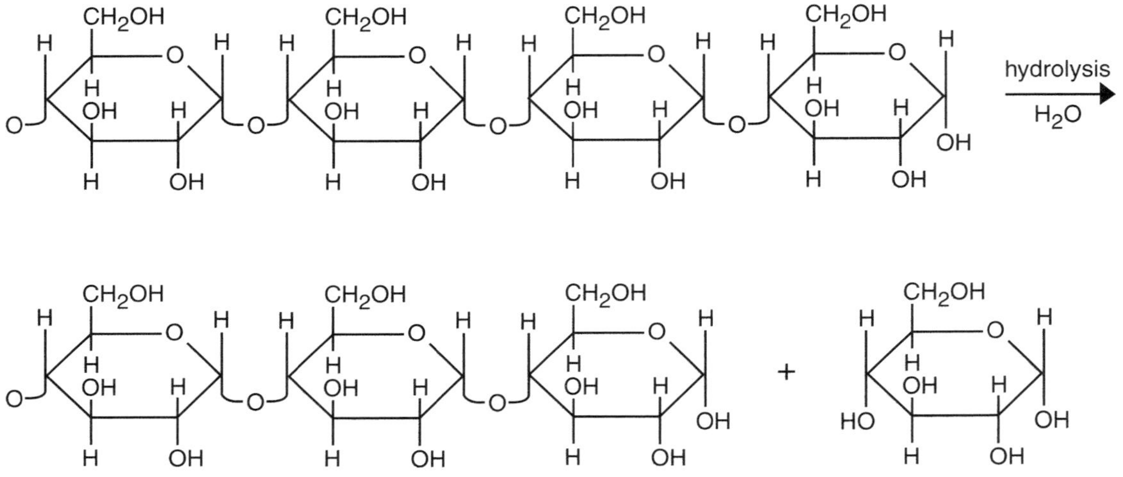

Fig. 1-6. Hydrolysis of starch to glucose units.

further described in Chapter 3. The mixture of glucose and fructose, the end products of the reaction, is also called *invert sugar*.

In a somewhat slower reaction, starch is also converted to glucose by our digestive systems. It takes a combination of acid and enzymes to do the job, but the overall reaction is fairly simple (Fig. 1-6). This reaction is also the basis for the large-scale commercial production of corn syrups and corn sugar (dextrose or glucose) from starch, the details of which are discussed in Chapter 3.

ISOMERIZATION

A specific reaction of importance in sweeteners is the conversion of D-glucose to D-fructose shown in Figure 1-7. The reaction involves an intermediate known as the "eno" form of the sugar. This reaction can be catalyzed by using a base such as sodium hydroxide, in which case

Invert sugar—A sugar consisting of equal parts of fructose and glucose that is made by the enzymatic breakdown of sucrose.

an equilibrium is set up between D-glucose, D-mannose, and D-fructose. To shift the equilibrium in favor of conversion to D-fructose, an enzyme called glucose isomerase is employed as a catalyst. This is the basis for the production of high-fructose corn syrup and crystalline fructose, described in more detail in Chapter 3.

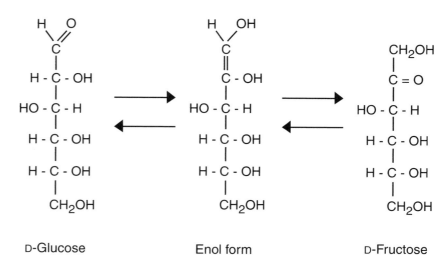

Fig. 1-7. Conversion of D-glucose to D-fructose through an enol intermediate.

REDUCTION

In carbohydrate chemistry, "reduction" refers to the conversion of aldehyde or ketone groups to hydroxyl groups (-CHOH). It is the basis for the production of *sugar alcohols* or *polyols* from sugars. The conversion of D-glucose to D-sorbitol, shown in Figure 1-8, is usually accomplished by the reaction of a sugar with gaseous hydrogen using a metal catalyst such as nickel or palladium. This reaction is also the basis for the production of other sugar alcohols, such as mannitol from mannose and maltitol from maltose.

OXIDATION

The oxidation reaction of sugars is important in the determination of *reducing sugars*. Reducing sugars are those sugars in which the carbonyl group can react to form a carboxylic acid group (Fig. 1-9). They

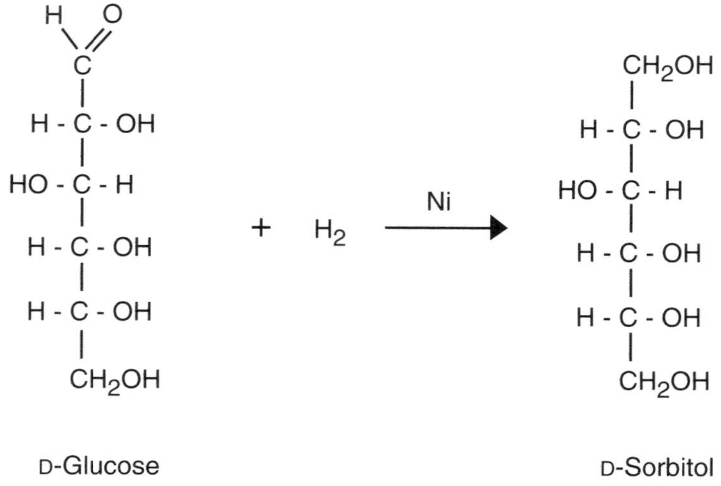

Fig. 1-8. Reduction of D-glucose to D-sorbitol.

Sugar alcohol or polyol— A compound derived by the reduction of sugar (in either the aldo or keto form), e.g., D-glucose to D-sorbitol.

Reducing sugar—A sugar molecule in which the carbonyl group can react to form a carboxylic acid group. The sugar can undergo nonenzymatic browning.

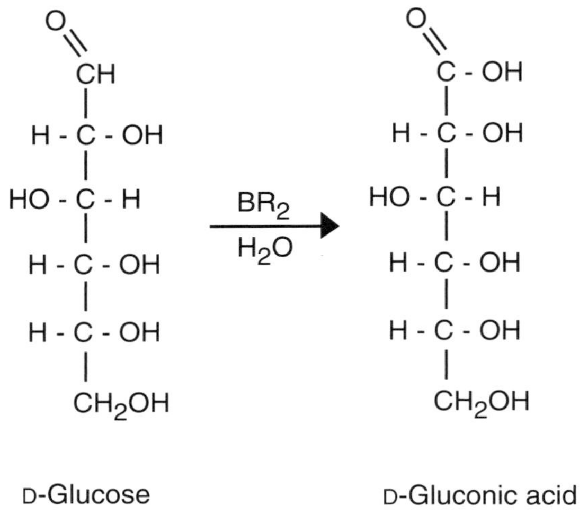

Fig. 1-9. Oxidation of D-glucose to D-gluconic acid.

Isomerization—Conversion of a molecule from one isomeric form to another. An **isomer** is a molecule with the identical components (number of carbons, hydrogens, etc.) as another molecule but a different structural makeup.

are able to react with compounds such as amino acids and proteins, which is important in the Maillard browning reaction, discussed below. The presence of reducing sugars is determined by using Fehling's reagent, alkaline cupric tartrate. It is important to note that sucrose, the most widely used sweetener, is a nonreducing sugar.

THERMAL DEGRADATION

Sugar molecules are stable under normal conditions. However, when heat is added, several reactions can occur, including degradation to other compounds, caramelization, and reaction with other food components such as amino acids or proteins. These important reactions involve a change in the sugar's structure and therefore a change in the sweetening power it will provide in the food system.

The formation of a thermal degradation compound depends upon the initial sweetener present, the pH of the environment, and the severity of the heat treatment. Most thermal degradations also involve the loss of a water molecule from the sugar component and therefore are also often called dehydration reactions. The first step in the reaction involves the *isomerization* of the sugar from the aldose form to the ketose form. Subsequent steps then lead to further degradation products. Pentoses that undergo dehydration

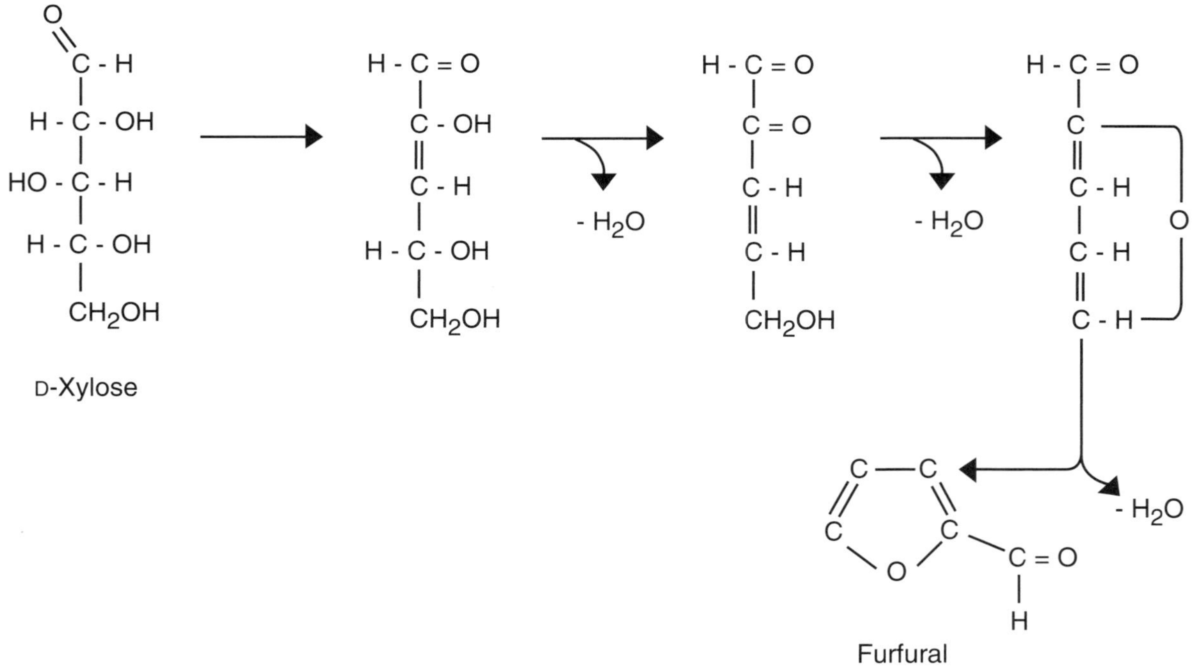

Fig. 1-10. Formation of furfural from a pentose.

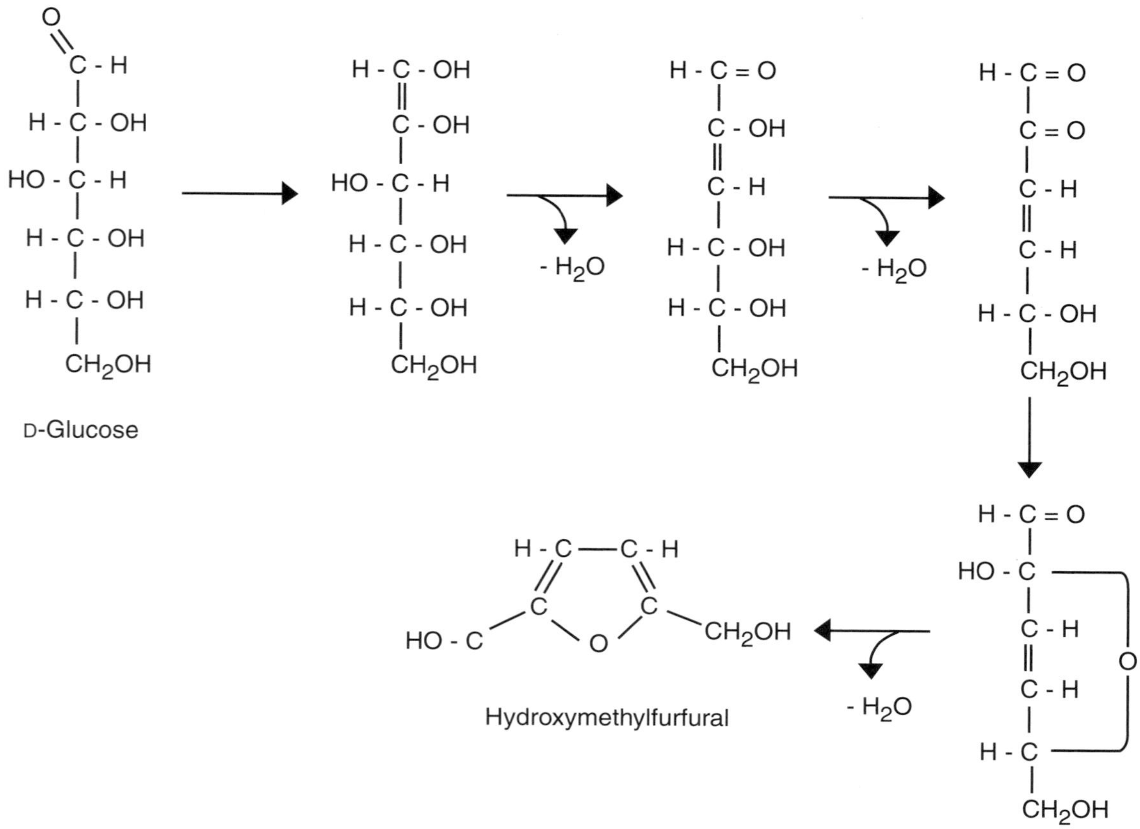

Fig. 1-11. Formation of hydroxymethylfurfural from a hexose.

with heat form furfurals (Fig. 1-10), while hexoses form hydroxymethylfurfural (Fig. 1-11). These reactions result in a loss of sweetness. Further thermal degradation of these products forms compounds such as formic acid or pyruvic acid.

Caramelization reactions involve the formation of brown pigments. The caramelization reactions can be tailored to pigment formation or aroma and flavor compound formation by manipulating the processing conditions of pH, temperature, and time.

Maillard browning. The reaction of sugars with amino acids or proteins, called *Maillard browning* (Fig. 1-12) or *nonenzymatic browning,* is very important in foods. Accelerated by heat, it is responsible for the browning and complex flavor development in foods that undergo heat treatments such as baking. In order for the reaction to take place, the presence of a carbonyl group (in this case from sugars) and an amine group (from amino acids or proteins) is necessary. The carbonyl group must be in its reactive form; that is, the sugar must be a reducing sugar. Since sucrose is a nonreducing sugar, it is unable to undergo Maillard browning. However, conditions such as pH, temperature, and the presence of enzymes can cause hydrolysis of the su-

Maillard browning—The browning of foods that occurs over time with high temperature. Also called **nonenzymatic browning** to distinguish it from the browning caused by an enzymatic reaction.

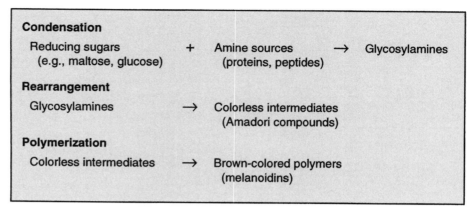

Fig. 1-12. Simplified Maillard reactions.

crose to its component sugars, fructose and glucose. Both of these are reducing sugars and hence can undergo Maillard browning.

Control of the Maillard browning reaction is based on the chemistry of the reaction. In the presence of reducing sugars and amino acids or proteins, increasing the temperature, increasing pH, and lowering the water activity all increase the rate of the reaction.

Supplemental Reading

1. El Khadem, H. S. 1988. *Carbohydrate Chemistry: Monosaccharides and Their Oligomers.* Academic Press, San Diego, CA.
2. Lineback, D. R., and Inglett, G. E. 1988. *Food Carbohydrates.* AVI, Westport, CT.

Sweetness as a Sensory Property

by Robert S. Shallenberger

Sweetness, saltiness, sourness, and bitterness are the primary tastes assigned to the assessment of food character and/or palatability. Taken together, they make up one of the so-called "chemical senses." The classification of taste as a chemical sense indicates that tastes are caused by chemical substances and implies that different tastes are probably caused by different chemical reactions. Also related to the sense of taste are olfaction (the sense of smell), astringency, and the senses of heat and cold.

In the pursuit of food, early humans probably used the sensation of sweetness to recognize wholesome food sources, since no toxic substances with an apparent sweet taste occur naturally. However, when mankind developed the ability to synthesize new chemicals, the general relationship between sweetness and wholesomeness was lost. Sweetness can now be found in all classes of chemical substances, some of which are known to be harmful (chloroform), cumulatively toxic (lead acetate, or "sugar of lead"), or incredibly toxic (beryllium chloride, or "glucinium").

Definition of Sweetness

The four primary tastes cannot be strictly defined. They are sensations that can be experienced only by demonstration. When placed upon the tongue, sucrose crystals and honey syrup taste sweet; crystals of sodium chloride taste salty; vinegar (acetic acid) tastes sour; and quinine tastes bitter. In other words, the basic tastes are the sensory experiences that result from the actual tasting of substances. If the "definition" or assessment of a taste is restricted to one specific substance, such as the sweetness of sucrose, no other substance can possibly have the same complete set of sweetness attributes. The complete set of taste attributes includes a compound's "taste profile" and "taste spectrum," no two of which are the same.

TASTE PROFILES

The taste profile for a substance encompasses the impact or onset time, the magnitude of the sensation at a given concentration, the

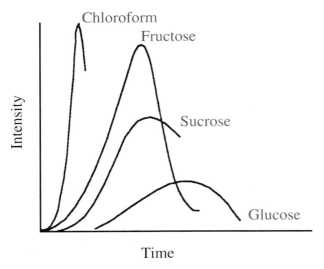

Fig. 2-1. Partial sweetness taste profiles for chloroform and common sugars.

temporal persistence characteristics of the sensation, the presence or absence of other tastes, and the presence or absence of tactile attributes such as "body" or "mouthfeel." Each chemical substance has a characteristic taste profile. A partial taste profile is obtained when the taste intensity of a substance is plotted against time.

For example, when a drop of chloroform is placed upon the tongue, the sweetness sensation occurs immediately, rapidly rises to a maximum, but then rapidly diminishes. When a drop of sugar solution is placed upon the tongue, the sweetness sensation gradually increases to a maximum and then decreases. All sugars have different sweetness profiles, as shown in Figure 2-1. The significance of relative sweetness scores is discussed later in this chapter.

TASTE SPECTRA

All standard taste substances are more or less capable of eliciting all four primary tastes. Also, the taste of a substance may vary. In dilute solution, sucrose tastes sour, and while saccharin is predominately sweet, it has an intrinsic bitter aftertaste. A taste "spectrum" is obtained by assigning the psychological assessment of the taste of a substance to all four of the potential taste attributes. An approximate taste spectrum for sucrose is shown in Figure 2-2.

The formulation of a taste spectrum for a substance is obviously a difficult task involving a taste panel, and such spectra must be interpreted carefully. The distinction between bitterness and sourness is known to be psychologically confusing to some people, and "taste blindness" to some substances may exist. About 25% of the population does not seem able to detect the bitter taste of phenylthiocarbamide, and rare cases of sweet taste blindness have also been reported. To some people, the taste of monosodium glutamate is unique and special ("umami," or tasty substance), but to others it tastes merely salty.

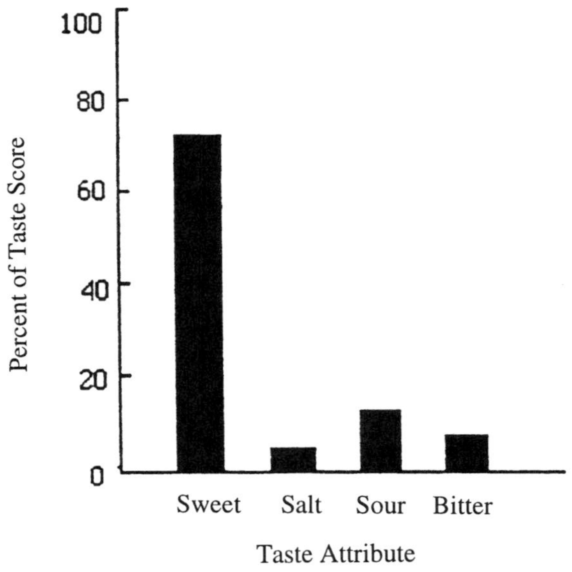

PERCEIVED SWEETNESS

When the intensity of a taste attribute is plotted against a wide range of concentrations, the response to the taste intensity of a substance describes a sigmoid function (Fig. 2-3).

Fig. 2-2. Taste spectrum for sucrose.

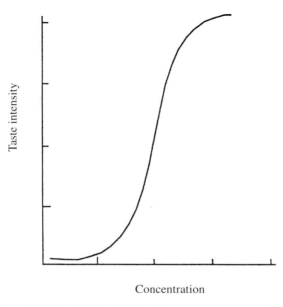

Fig. 2-3. Typical taste intensity response to a wide range of sample concentrations.

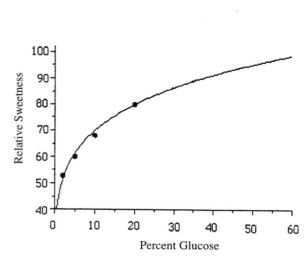

Fig. 2-4. Relative sweetness of glucose with increasing concentration. Sucrose is the reference compound.

An inductive effect, or accelerating segment of the curve, is observed as the concentration approaches the recognition threshold. This is followed by a near-linear function. As mass-action effects take hold at the inflection point in the sigmoid curve, the plot demonstrates decelerating characteristics. The course that the perceived taste takes with increasing concentration is known as "perceived sweetness." Other terms that have been employed are "absolute sweetness" and "intrinsic sweetness." For the perceived sweetness of sucrose over the practical use range of 1–27% (w/v), the relationship between sweetness and concentration is approximately linear.

RELATIVE SWEETNESS

Although no two sweeteners have the same sweetness profile and taste spectrum, single-figure "relative sweetness" scores are often used to compare substances. The reference standard is usually sucrose. Apparently, a person is capable of integrating the total response from a compound's taste profile and taste spectrum and then assigning a single (unidimensional) value or score to the overall (multidimensional) sensation.

Because the single sweetness scores are an integrated assessment of a compound's taste profile and spectrum, relative scores vary with many factors, such as the concentration of the reference sample. In general, substances with less sweetness than sucrose, on an equimolar basis, appear to increase in relative sweetness with increasing concentration. This is shown for glucose in Figure 2-4.

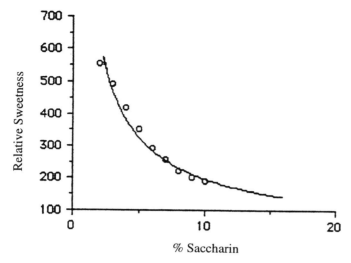

Fig. 2-5. Decrease in relative sweetness of saccharin with increasing concentration. Sucrose is the reference compound.

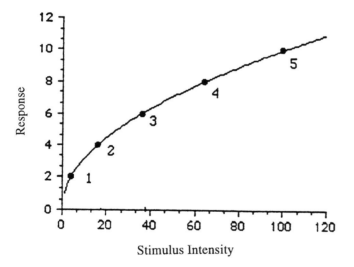

Fig. 2-6. Exponential increase in stimulus intensity required to elicit an arithmetic increase in a sensory response.

For substances with greater relative sweetness than the reference compound (sucrose), relative sweetness declines rapidly with increasing concentration. It takes more and more high-potency sweetener to equal the sweetness imparted by sucrose at higher concentrations; hence, the relative sweetness of the high-potency substance appears to decline. This trend is shown for saccharin versus sucrose in Figure 2-5.

The variation in relative sweetness results from several factors. Foremost among them are the dynamics of sensory perception, i.e., the psychophysical laws. The first principle of those laws states that, with increasing increments of a sensory stimulus, larger and larger increments of stimulus intensity are required to elicit a measurable difference (a "just noticeable difference," or JND) between the two stimuli. In the example shown in Figure 2-6, a 10-unit increase between stimulus intensity parameters 1 and 2 yields a two-unit increase in the sensory response. To elicit the same JND (two units), a 30-unit increase in stimulus intensity is required between stimulus intensity units 4 and 5. Thus, the *relative* sweetness of a high-potency sweetener must necessarily decrease with increasing concentrations of the sweetener.

A second reason for relative sweetness variation is the alteration in the chemical composition of the sweetener at different temperatures. When solutions of the sugar fructose are subjected to changes in temperature, a shift occurs among the several structural forms of the sugar that are present. The different structural forms are easily interconverted isomers, which do not have equal sweetness.

A third reason for differences in relative sweetness is adaptation. When a person is repeatedly exposed to the taste of a substance, the perceived taste intensity decreases from its initial level. For example, the first bite of grapefruit in the morning may seem extremely sour, but subsequent bites seem less and less sour, as a balance between sweetness and sourness is reached. Both adaptation and recovery from adaptation follow first-order reaction kinetics and are therefore probably manifested in simple chemical equilibrium kinetics. Since adaptation rate is proportional to sweetness potency, the phenomenon plays a role in relative sweetness variation.

Chemical Structure and Sweetness

Early in the 20th century, scientists recognized that, in order to taste sweet, a compound had to possess special functional groups. These groups (e.g., OH, NH, and NO_2) usually occurred in pairs, and the pair of groups was named a "glucogene." Subsequently, different chemical functions were assigned to the component parts of the glucogene. Based on dye chemical theory, one was described as an "auxogluc" (usually a hydrogen atom) and the other as a "glucophore." In 1967, Shallenberger and Acree (1) pointed out that all compounds that possess sweet taste possess a bipartite system capable of concerted hydrogen bonding. In essence, the auxogluc was seen to function as a proton-donating group and the glucophore as a proton acceptor. The active pair is now described as a "glycophore", and is believed to be common to all compounds that taste sweet.

Glycophores are dipolar in nature. That is, they have opposite forces (attractive and opposing) at opposite ends of the molecule. The proton-donor group is often termed "AH," and the proton acceptor is termed "B." In Figure 2-7, the proton donors and acceptors are also labeled "e" and "n" because they can also be construed to be *electrophilic* and *nucleophilic*, respectively.

The establishment of AH,B as the factor common to all sweet-tasting substances led to description of the mechanism for the initial chemistry of sweet taste shown in Figure 2-7. Among the most serious criticisms of the thesis was the fact that not all compounds that possess a hydrogen bonding (AH,B) system taste sweet (some are actually bitter). As it now turns out, this criticism can be neatly resolved by the application of symmetry principles. That is, for sweetness, it is believed that the charge on the *dipole* of the sweet substance must be *bilaterally symmetrical*, i.e., of opposite sign but of nearly equal antipodal activity. When the charge is unbalanced or when only one component of a dipole (a monopole) is present, there is a propensity for organic substances to taste bitter.

While the Greek word elements gluc- and glyc- both indicate sweetness, glyc- is an established generic structural term in carbohydrate chemistry, whereas gluc- has a specific structural connotation indicating the gluco- or glucose configuration.

Electrophilic—Having an affinity for negative charges.

Nucleophilic—Having an affinity for positive charges.

Dipole (dipolar)—A group of atoms having equal electric charges of opposite sign that are separated by a finite distance.

Bilateral symmetry—The balanced combination of left and right forms.

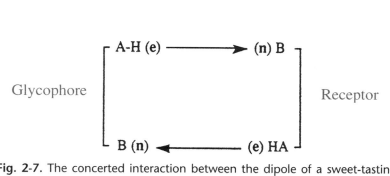

Fig. 2-7. The concerted interaction between the dipole of a sweet-tasting organic compound and a dipolar group of the taste receptor.

Upon examining the structural features of the D-amino acids, Kier (2) noted that a group high in electron density occurred in the same position with respect to the *zwitterionic* functions (AH,B) of the amino acids. This group was originally designated "X." In all cases, the tripartite arrangement of the groups AH, B, and X formed a scalene arrangement with an AH-to-B distance of 2.6 Å, an AH-to-X distance of 3.5 Å, and a B-to-X distance of 5.5 Å. Although Kier noted that the third component functions primarily as a "dispersion" site, others recognized that it possesses multiple functions that include a *hydrophobic* function and an electron-withdrawing (inductive) capacity. It may also serve as a hydrophobic probe that "steers" a sweetener to and aligns it with the receptor site. Thus, the third glycophoric component was designated γ. The letter "C" might have been used for the third glycophore component (hence making an ABC glycphoric unit), but γ (the Greek letter gamma) seems to better indicate its lipoid ("fatty" or "greasy") and consequently hydrophobic character.

Glycophores are found in compounds that range in size and chemical classification from simple halogenated hydrocarbons (e.g., chloroform) to sweet-tasting proteins. Several examples of AH,B and γ for selected compounds are shown in Figure 2-8.

A second criticism of the AH,B theory of sweetness was that the L-series of sugars would, in apparent contradiction to known stereochemical principles, need to taste sweet. A major feature of a third multifunctional site in the composition of the glycophore is that it serves as the basis for resolving this chiral (handedness) anomaly with respect to the sweetness properties of enantiomeric (mirror image) sugars and amino acids. Resolution of the anomaly serves, in turn, to indicate the method whereby sweeteners interact with a receptor (3). Briefly, early in the history of sweetness and structure studies, scientists observed that crystalline D-asparagine, unlike the natural L-form, tasted sweet, and they concluded that the sweetness receptor must therefore possesses handedness. Subsequently, it was established that the relationship between amino acid chirality and taste was not straightforward. In some cases, both enantiomeric forms were sweet (e.g., alanine, serine, and threonine), but in others, the D-form was tasteless while the L-form was sweet (e.g., proline and hydroxy proline). Nevertheless, it came to be tacitly assumed that since the sweet taste receptor could distinguish between D- and L-asparagine, it must also be able to distinguish between the D- and L-sugars, and since the D-forms of the sugars are naturally occurring and taste sweet, the L-forms must be tasteless. This tacit assumption was not true, but the fact that L-sugars do indeed taste sweet had not yet been discovered. It is now known that the three-dimensional attributes of substances are not operative in taste chemistry and that the tripartite two-dimensional attributes of the faces of *tastants* determine the taste of enantiomers. It was through such considerations that the fact that the L-sugars taste sweet was discovered.

When the third component (γ) of a glycophore possesses strong electron-withdrawing power, it fills the requirements for high-

Zwitterionic—Having both positive and negative charges.

Hydrophobic bond—An interaction, i.e., attraction, between two apolar groups in a polar (aqueous) environment.

Tastant—A substance capable of eliciting taste.

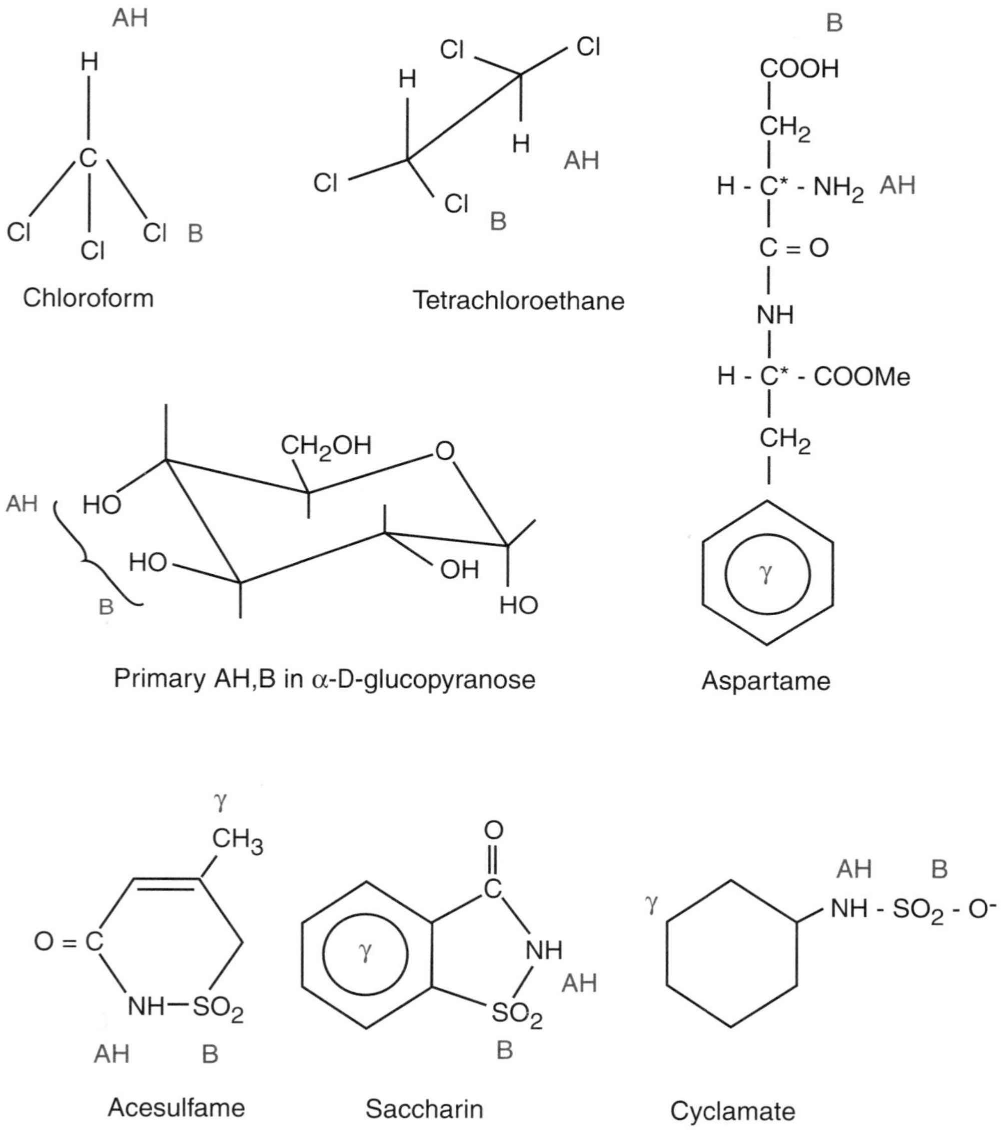

Fig. 2-8. Glycophores and their AH, B, and γ designations.

potency sweetness. The fact that high-potency tastes are found only among the sweet and bitter organic substances indicates that this phenomenon is also related to structure and composition. In the final analysis, high-potency sweetness and bitterness occur whenever, through intrinsic structure or by group substitutions, a strong electron-withdrawing effect occurs that serves to promote or enhance the activity of AH,B. Some structures with this effect are cyclic organic compounds (e.g., saccharin, cyclamate, and aspartame). Some electron-withdrawing groups are NO_2, Cl, and CN.

References

1. Shallenberger, R. S., and Acree, T. E. 1967. Nature 216:480.
2. Kier, L. B. 1972. J. Pharm. Sci. 61:1394.
3. Shallenberger, R. S. 1993. *Taste Chemistry.* Blackie Academic & Professional, London.

Supplemental Reading

Moncrieff, R. W. 1967. *The Chemical Senses,* 3rd ed. Leonard Hill, London.

Production and Description

Many different types of sweeteners are available for product development today. It is important to have an understanding of what each of these sweeteners is in order to understand the functions that it will serve in the food system. The production processes for and types of major carbohydrate-based sweeteners are described below.

Sucrose-Based Sweeteners

PRODUCTION

Although other plant sources such as sorghum, date, and palm are available, the industrial production of sucrose is based exclusively on sugarcane and sugar beet processing. Sugarcane is a tropical plant, while the sugar beet is grown in cooler climates. Contrary to popular belief, there are no known structural, chemical, or physical differences between the sucrose obtained from sugarcane and that obtained from sugar beets. However, the processing is different. Processing of cane sugar is normally a two-step operation. It starts in raw-sugar processing plants located in tropical or subtropical regions close to the sugarcane fields. The *raw sugar* is then transported to refineries (built near major areas of consumption) and refined into high-purity products. Beet-growing regions are close to major consumption areas; consequently, beet sugar processing and refining are done at the same location.

Sucrose is contained in cane and beet cells. The first processing step (Fig. 3-1) is the extraction of sucrose from these plant tissues. In the cane mills, extraction is done by a series of roller presses, while beet sugar factories use countercurrent extraction with hot water (a diffusion process).

Solutions obtained by these processes have a purity of 84–86% with retained dry solids (RDS) of 14–16% and contain various amounts of impurities from plant tissues. Some of the nonsugar substances can be eliminated by simple mechanical screening, while others must be flocked and separated by settling and filtering. This part of the processing, called "juice purification," is the most sensitive and demanding part of sucrose manufacturing. In some factories, it is done with lime-carbon dioxide juice-purification systems, while in most refineries it is done with the lime-phosphoric acid flotation process. Further improvements in quality are achieved by activated

Raw sugar—Sugar that has not undergone the refining process.

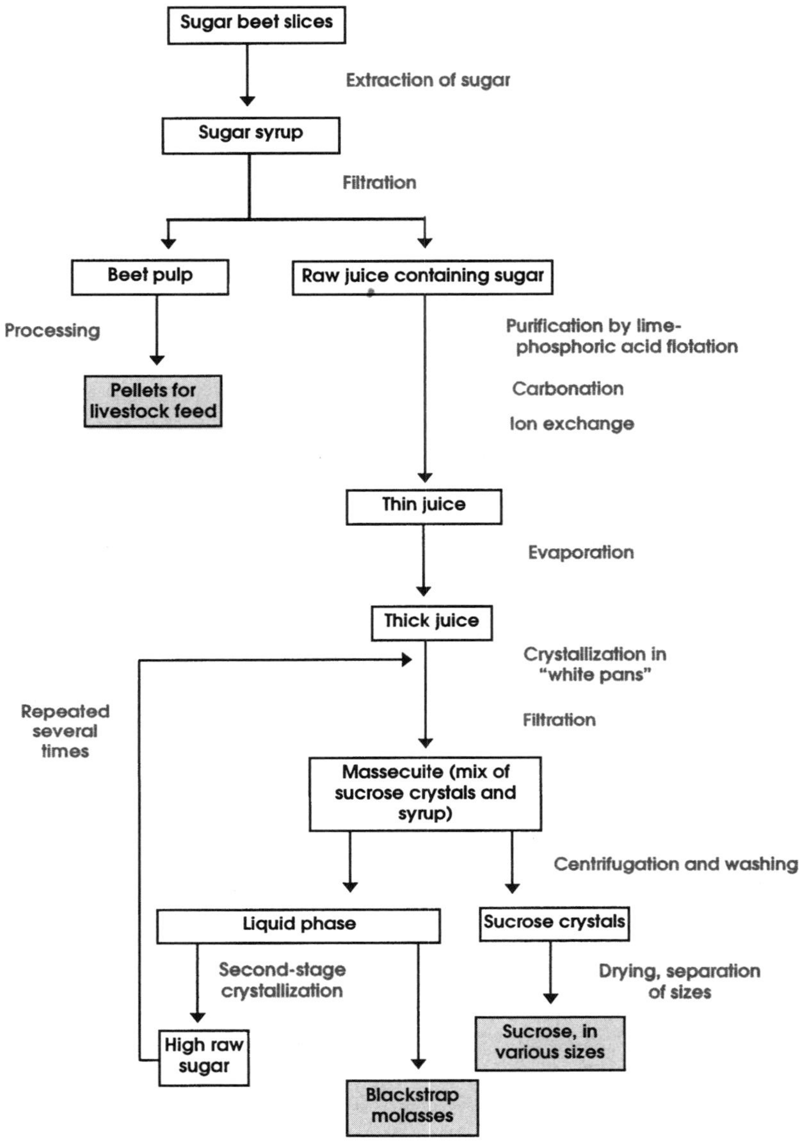

Fig. 3-1. Production process for sucrose.

carbon and ion exchange treatments, which decolorize and demineralize the juice. During this phase, the juice (in both beet and cane plants) has a purity of 91–92% and RDS level of 12–15%.

The next step is evaporation. A large volume of water is evaporated from the juice in multiple-stage evaporators. The goal is to obtain juice with RDS of 65–71%. To achieve this goal, the evaporators must separate 3,000–15,000 tons of water per day (depending on the processing capacity). Multiple-stage evaporation enables modern plants to accomplish this enormous task with high efficiency and economy. The high-density syrup achieved by evaporation is called the "thick juice" (or "standard liquor" if recovery sugar is dissolved in the thick juice).

The thick juice then undergoes crystallization in vacuum pans (also called "white pans"). Here, under controlled conditions (involving, e.g., pressure, boiling temperature, density, purity, and viscosity), the dissolved sucrose is transformed into crystals. The material formed in the vacuum pan is called the *massecuite*. In the massecuite, the sucrose crystals are mixed with a high-density liquid called "mother liquor."

The next step is to separate the crystals from the liquid phase by centrifugation (1,000–2,500 x g). The wet sucrose is then dried in a stream of hot air in rotating drums, cooled, classified on vibrating screens, and sent to sugar packaging or bulk storage. The remaining liquid phase is then further processed in second-stage crystallization, which is identical to the first-stage process. The sugar obtained from this stage, which is lower in purity and is called "high raw sugar," is redissolved in water or juice and sent back to the "white sugar" process. The goals of the multiple-stage process are maximum sugar exhaustion from syrups and the production of premium quality sugar. Depending on the particular factory, this multiple-stage crystallization is repeated two to seven times. In principle, the low-purity syrups that result from crystal separation are sent downstream, while the crystals of the various stages are remelted and moved upstream in the production line. Eventually, two products leave the crystallization station: high-purity, crystallized sucrose and the low-purity liquid phase, called "blackstrap *molasses*."

PRODUCT TYPES

Sucrose products manufactured and sold in the United States are generally classified into four categories: granulated sugar, liquid sugars, brown sugars, and specialty products. Some sucrose products are shown in Box 3-1.

Box 3-1. Sucrose Products

High-purity sucrose: From cane refineries and beet sugar plants, 99.90–99.95% purity, 0.02–0.04% moisture content

Brown and soft sugars: From cane refineries and beet sugar plants, 92.00–98.00% purity, 0.4–3.5% moisture content

Raw sugar: From cane raw sugar mills, 96.50–97.50% purity, 0.5–0.7% moisture content

Blackstrap molasses: From cane refineries, 38.00–45.00% purity, 82.0–88.0% retained dry solids

Raw molasses: From beet sugar factories, 56.00–62.00% purity, 82.0–86.0% retained dry solids

---

Massecuite—A dense mixture of sugar crystals and syrup that is an intermediate product in the manufacture of sugar.

Molasses—A thick, brown, uncrystallized syrup produced during the refining of sucrose.

TABLE 3-1. Selected Properties of Granulated Sugars

Property	Coarse	Sanding	Extra Fine	Fruit	Bakers' Special	Powdered 6X	Powdered 10X
Color, ICU	20–35	20–35	25–50	25–50	25–50	25–50	25–50
Ash, % (max.)	0.015	0.015	0.02	0.03	0.03	0.03	0.03
Moisture, % (max.)	0.04	0.04	0.05	0.05	0.05	0.5	0.5
Starch, %	...	...	...	...	...	2.5–3.5	2.5–3.5

Granulated sugar. Granulated sugar is a pure, crystalline material with more than 99.8% (dry basis) sucrose content. It is produced in a controlled crystallization process in vacuum pans and is classified by the percentage of crystals retained on a particular standard U.S. mesh screen (Table B-1 in Appendix B). Some properties of various types of granulated sugar are shown in Table 3-1.

Liquid sugars. Essentially, liquid sugar is granulated sugar dissolved in pure water. It is produced at minimum concentrations of 67% solids and 99.5% sucrose. The product can be used wherever a dissolved granulated sugar product is needed. All liquid sugars are sold at the highest concentration of solids at which the sugar will remain in solution at 21°C (70°F). Some properties are shown in Table 3-2 and some in Table B-2 in Appendix B.

Brown sugars. Brown sugar is basically a fine-grain, granulated sugar covered with a thin layer of cane syrup from the cane refinery recovery scheme. The sugars are classified according to the processing method: soft brown sugar, coated brown sugar, and free-flowing brown sugar (Table 3-3 and Table B-3 in Appendix B). Brown sugars are used to develop a rich, molasses-type flavor.

Specialty products. Specialty sugars can be defined as sugar products made to meet a specific need, e.g., cube sugar, fondant sugar, cocrystallized sugar, flavored sugar, and agglomerated sugars.

TABLE 3-2. Selected Properties of Liquid Sugars

Property	Liquid Sucrose	Amber Sucrose	Liquid Invert	Total Invert
Solids, %	67.0–67.9	67.0–67.7	76–77	71.5–73.5
Color, ICU	≤35	≤200	≤35	≤40
Ash, %	≤0.04	≤0.15	≤0.06	≤0.09
pH	6.7–8.5	6.5–8.5	4.5–5.5	3.5–4.5

TABLE 3-3. Selected Properties of Brown Sugars

Property	Soft Brown Light	Soft Brown Dark	Coated Brown Light	Coated Brown Dark	Free–Flowing Granulated	Free–Flowing Powdered
Ash, %	1–2	1–2.5	0.3–1	0.3–1	1–2	1–2
Moisture, %	2–3.5	2–3.5	1–2.5	1–2.5	0.4–0.9	0.4–0.9
Color, ICU	3,000–6,000	7,000–11,000	3,000–6,000	7,000–11,000	6,000–8,000	6,000–8,000
Color, reflectance	40–60	25–35	...	...	...	...

Molasses. The beet sugar industry defines molasses as the heavy, viscous liquid separated from the final low-grade massecuite from which no further sugar can be crystallized by the usual methods. In the cane refineries, the same heavy liquid is called "blackstrap molasses." Another type of molasses, "edible molasses," is a clear, light brown, 80° Brix syrup of 45–50% purity that is generally sold in bulk for blending. The old-fashioned New Orleans molasses was the by-product of open-kettle boiling. The clarified juice was boiled in an open kettle until sucrose crystals were formed. After the sucrose was separated by centrifugation, the mother liquid was sold as edible molasses. A third type, "high-test molasses," is a heavy, partially inverted cane syrup. All edible molasses are characterized by dark brown color, distinctive flavor, and high densities (RDS = 85%).

Invert sugar. Invert sugar is produced from sucrose to yield glucose and fructose through the use of the enzyme invertase or by hydrolysis or cation exchange resins. The final product is available as a syrup. Invert sugar is an important raw material in the production of sugar alcohols such as sorbitol and mannitol.

Starch-Based Sweeteners

Many different sweeteners are derived from starch. The starch can come from rice, wheat, oats, and potatoes, but in the United States, the primary source is corn because of its availability and relatively low cost compared with other starch sources. The production methods for the various corn-based sweeteners are fairly similar or at least interrelated.

DEXTROSE-BASED PRODUCTS

Dextrose equivalent (DE) is a measure of the extent of starch hydrolysis. It is determined by measuring the amount of reducing sugars in a sample relative to dextrose. The DE for dextrose is 100, representing 100% hydrolysis. Starch-based sweeteners that are not fully hydrolyzed have a DE of less than 100. The lower the extent of hydrolysis, the lower the DE. Dextrose syrups, which have a high DE (95 and greater), are often referred to as liquid dextrose.

Maltodextrins. The least hydrolyzed starch-based products are maltodextrins. These materials are made by either a one- or two-stage process (Fig. 3-2). In the one-stage process, normally used with waxy corn starch, a starch slurry containing 30–35% solids (Fig. 3-2, step 1) is treated with an α-amylase, heated in a jet cooker (step 2), which produces a very low-DE (0–5) hydrolysate, and held (step 3) until the desired DE is obtained. Enzyme activity is terminated; pH is adjusted to 4.0–5.0; and the hydrolysate is refined by the standard techniques used for corn sweeteners (steps 6–12).

The two-stage process is normally used with regular dent varieties of corn starch. It employs either an acid or an enzyme in a high-

Dextrose equivalent (DE)— A measure of the percentage of glucosidic bonds hydrolyzed. Dextrose has a DE of 100.

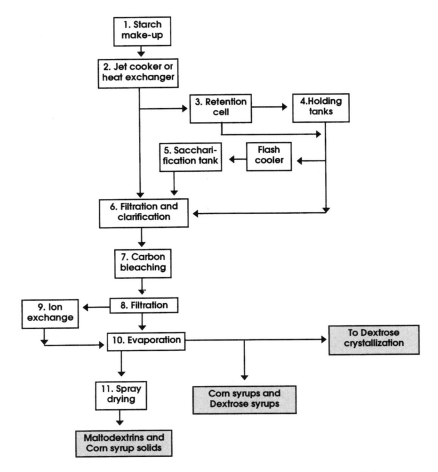

Fig. 3-2. Conversion of starch to maltodextrins, corn syrups, and dextrose.

temperature treatment (above 100°C) through a jet cooker (step 2). The hydrolysate is cooled and then passed through a second stage (step 4) involving enzyme hydrolysis with α-amylase to the desired DE. After the DE is obtained, the hydrolysate is refined as in the one-stage process.

From either of these two processes, maltodextrins are obtained as white, spray-dried solids containing 4–6% moisture. They are soluble in water and yield colorless, very bland solutions. Their characteristics are summarized in Table B-4 in Appendix B. Maltodextrins contain very low levels of dextrose and maltose and are therefore not sweet.

Corn syrups and corn syrup solids. Corn syrups are usually produced by a two-stage process (1). The first step involves the acid hydrolysis of starch to about 42 DE as shown in steps 1–3 of Figure 3-2. If a higher DE is required, the starch is kept in a holding tank (step 4) while a second-stage hydrolysis with enzymes is employed until the desired DE is obtained. The possible acid-enzyme and enzyme-enzyme combinations are endless, and the precise combination used depends upon the desired properties and composition of a syrup for a particular application. Finished syrups are produced by standard refining and evaporation procedures (steps 6–10).

The most common corn syrups have DE levels of 42 and 62, although syrups in the 24- to 82-DE range have been produced. In addition, specialty brewers' syrups are produced for specialty applications. Corn syrups are colorless, viscous liquids containing 74–84% solids. Characteristics of the standard syrups are summarized in Table B-5 in Appendix B.

Corn syrup solids are available at DE levels that range from 20 to 48. The process for manufacturing corn syrup solids depends on the DE level. Products with a higher DE, ranging from 30 to 48, are made from corn syrups that have been further dried and crystallized. Corn syrup solids with a DE of 20–30 are made by a process similar to that

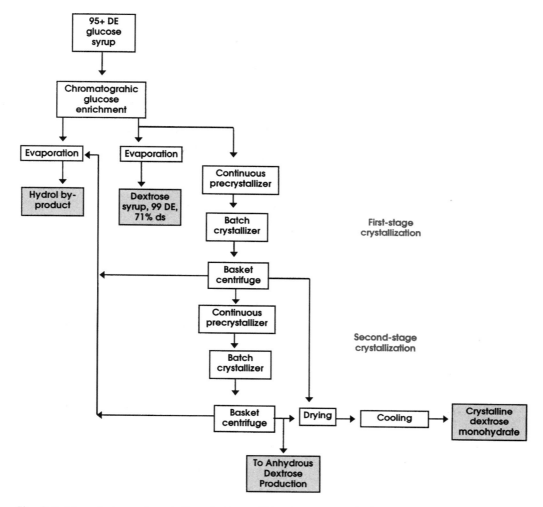

Fig. 3-3. Manufacture of crystalline dextrose. DE = dextrose equivalent.

used to make maltodextrins. Characteristics of corn syrup solids are shown in Table B-5 in Appendix B. All corn syrup solids have sweetness levels similar to those of their corresponding corn syrups. They have an advantage over corn syrups in that they can be used in low-moisture systems or where dry ingredients are required.

Dextrose. Dextrose is the common name for D-glucose, where D-stands for *dextrorotary*. It is the product of the complete hydrolysis of starch by both acid and enzymes, having been treated with α-amylase (Fig. 3-2, step 2) and glucoamylase (step 5). The hydrolysate is first purified and then crystallized to yield dextrose monohydrate (Fig. 3-3). Further drying yields anhydrous dextrose.

Dextrose syrups can be either 95 or 99 DE (Table B-6 in Appendix B). Crystalline products with a DE range of 93–99, a solids range of 91–99%, and pH of 4.5 are available in various granulations.

Dextrorotary—Describing a compound that can cause a plane of polarized light to rotate in a clockwise fashion (to the right). Compounds that cause polarized light to rotate in a counterclockwise direction (to the left) are termed "levorotary."

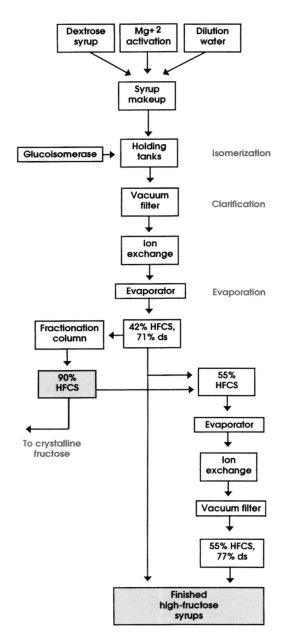

Fig. 3-4. Manufacture of high-fructose corn syrup (HFCS) from glucose syrup.

Hygroscopicity—The ability to attract and retain moisture.

FRUCTOSE-BASED PRODUCTS

High-fructose corn syrup. The various grades of high-fructose corn syrup (HFCS) are produced by the process outlined in Figure 3-4. High-DE (95+) dextrose syrup is first made according to the steps outlined in Figure 3-2. The hydrolysate is then passed over immobilized glucose isomerase, and the glucose is partially converted to fructose. This results in a syrup containing 42% fructose, 53% glucose, and 5% more-complex sugars. This 42% HFCS is passed over specially sized exclusion chromatographic columns, which increases the fructose in the syrup to 90%. The 90% syrup is blended with 42% syrup to give a 55% fructose syrup (55% fructose, 41% glucose, and 5% sugars that are more complex). The 42% fructose syrup is approximately as sweet as sucrose, while the 55% product is 1.2 times as sweet.

All the syrups may then be refined by standard techniques to give finished HFCSs. Part of the 42% syrup is concentrated and sold as 42% HFCS (71% dry substance [ds]). The 55% syrup is also concentrated and sold as 55% HFCS (77% ds) Some of the 90% syrup is used to make crystalline fructose (Fig. 3-5), and some is sold as 90% HFCS. Characteristics of the various types of HFCS are shown in Table 3-4.

Crystalline fructose. The exact process for making crystalline fructose is still somewhat proprietary, since there are few large-scale commercial manufacturers in the United States. However, an outline of the general process for making the product is shown in Figure 3-5. Details of a specific process employing a mixed-alcohol solvent are described in a U.S. patent (2).

In general, the process involves adding a solvent or mixed-solvent system to the 90% fructose syrup originating from the HFCS process. A nonaqueous solvent is apparently needed because of the very high level of *hygroscopicity* associated with fructose and the difficulties in crystallizing fructose from a water solution. This mixture is then evaporated to about 90% solids and fed to the crystallizers. The material is centrifuged, and the fructose crystals are dried, cooled, and packaged.

The physical properties of a crystalline fructose product are shown in Table 3-5. Most fructose is manufactured as the crystalline material, although some is sold as liquid fructose, an 80% solids syrup.

TABLE 3-4. Characteristics of Typical High-Fructose Corn Syrups

| Type (% Fructose) | Percent Solids | pH | Composition, % | | | Relative Sweetness[b] |
			Glucose	Fructose	Higher DP[a]	
42	71–80	4.0	53	42	5	0.9–1.0
55	77	3.5–4.0	41	55	4	1.0–1.2
90	80	4.0	7	90	3	1.4–1.6

[a] Degree of polymerization.
[b] Sweetness relative to sucrose at 1.0.

OTHER STARCH-BASED PRODUCTS

Maltose. Maltose is a disaccharide consisting of two molecules of glucose. Pure maltose is not produced on a commercial scale in the United States. It is usually imported from either France or Japan, where it is manufactured on a large scale and sold in syrup, powder, and crystalline forms in several grades of purity (3).

Maltose is the main component of high-maltose syrups, which are made by treating regular corn syrups or partial starch hydrolysates with β-amylase. Further treatment of these syrups with debranching enzymes (pullulanase and isoamylase) produces syrups containing 70–90% maltose, which is about the upper limit of maltose content. Characteristics of high-maltose corn syrups are shown in Table B-6 in Appendix B. Production of pure maltose can be accomplished with various fractionation techniques. Presently, cation-exchange chromatography and ultrafiltration are used to make high-purity maltose syrups on an industrial scale.

Maltose has many of the characteristics of sucrose but is only 30–40% as sweet. High-maltose syrups and crystalline maltose offer good stability, reduced color formation, and low hydroscopicity, all useful properties in some applications.

Crystalline maltose has two anomeric configurations, α and β. In solution, they equilibrate at an α-β ratio of 42:58. Slow crystallization from water yields three crystalline forms: β-maltose monohydrate, anhydrous α-maltose, and a complex containing both.

Malt syrup. Several malt products are used by the brewing industry in making beers and ales. The main products of interest as sweeteners are malt extract, malt syrup, and rice syrup. Malt extract is made from germinated barley (i.e., malt), which is allowed to become partially hydrolyzed in an aqueous system. It is then extracted, filtered, purified, and evaporated to give a 78–80% solids syrup.

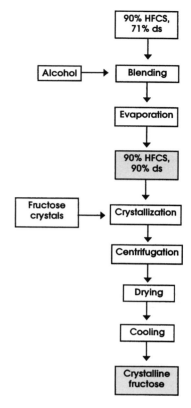

Fig. 3-5. Manufacture of crystalline fructose. HFCS = high-fructose corn syrup.

TABLE 3-5. Typical Properties of Crystalline Fructose

Property	
Molecular formula	$C_6H_{12}O_6$
Molecular weight	180.16
Chemical name	β-D-Fructopyranose
Melting point, °C	102–105
Caloric value, Kcal/g	3.8
Solubility, g/100 g	400
Relative sweetness[a]	1.2–1.6
Composition	
Fructose, %	99.5
Glucose, %	0.5
Moisture, %	0.05
Ash, %	0.1

[a] Relative to sucrose at 1.0.

Malt syrup is made by combining malt and corn grits (or meal), partially hydrolyzing the mixture, and then extracting, purifying, and evaporating the resulting syrup. Rice syrup is available from some of the manufacturers of malt products. It is made by combining malt with rice grits or flour by the same processing steps as used for malt syrup. Malt extract is basically barley syrup, malt syrup is barley and corn syrup, and rice syrup is barley and rice syrup.

The various products are termed either diastatic or nondiastatic, based on the enzyme activity. Diastatic products contain various levels of residual enzyme activity. The properties of several nondiastatic products are summarized in Table 3-6.

TABLE 3-6. Typical Properties of Nondiastatic Malt Extracts and Syrups[a]

Property	Extracts (all barley)		Malt Syrups (barley/corn)	
	Light	Dark	Light	Dark
Solids, %	78–80	78–80	78–80	78–80
Reducing sugars, as maltose, %	53–63	53–63	60–72	60–72
Protein, %	4.5–5.6	4.5–5.6	1.8–3.5	1.8–3.5
pH, at 10% solids	5.0–5.7	5.0–5.7	5.0–5.7	5.0–5.7
Color, Lovibond	100–300	250–425	80–150	225–400

[a] Data from (4).

Others

HONEY AND ARTIFICIAL HONEY

Honey. Honey is the substance made when the nectar and sweet deposits from plants are gathered, modified, and stored in the honeycomb by bees. The unique flavor of each lot of honey is attributable to the floral source from which the honey bees gathered the nectar—there are more than 300 such sources in the United States.

Honey is the only natural sweetener that needs no additional refining or processing to be utilized. However, honey is commonly heated to destroy yeasts and delay crystallization and is usually filtered or strained to remove extraneous material (Fig. 3-6). Honey is generally sold as a liquid but is also available in a thicker, opaque form known as cremed, spun, whipped, or churned honey. This form

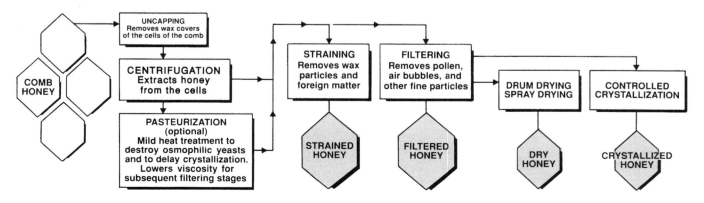

Fig. 3-6. Processing of honey. (Courtesy of the National Honey Board)

is made by closely controlling the moisture of the liquid honey, then heating it, seeding it with fine crystals, and whipping it. Dried honey is also commercially available.

Honey is an invert sugar composed mainly of fructose and glucose (Table 3-7) as well as small amounts of vitamins and minerals (Table B-7 in Appendix B). Its color ranges from nearly colorless to dark brown, varying with the mineral content. As a general rule, lighter-colored honey is milder in taste and darker-colored honey is stronger.

Honey is graded using a point system set up by the U.S. Department of Agriculture based on water content, flavor and aroma, and absence of defects. U.S. Grade A (90 points) and U.S. Grade B (80 points) contain <18.6% water; U.S. Grade C (70 points) contains ≥20% water.

TABLE 3-7. Average Composition of Honey[a,b]

Component	Percent
Moisture	17.1
Fructose	38.5
Glucose	31.0
Maltose	7.2
Sucrose	1.5
Higher sugars	4.2
Protein	0.5
Acids	0.6

[a] pH 3.9.
[b] Data from (5).

Artificial honey. Artificial honey is essentially invert sugar syrup or corn syrups that have been altered in appearance and flavor to mimic natural honey. It is composed of invert sugar, sucrose, water, ash, and a crystallization inhibitor, which is usually a hydrolyzed starch.

LACTOSE

During recent years, there has been considerable interest in the production (6) and utilization of lactose as a by-product from whey, which itself is a by-product of cheese manufacturing. For every pound of cheese produced, approximately 9 lb of whey is also obtained. Whey contains about 6% solids, of which 4.7% is lactose, 0.7% is protein, and 0.5% is minerals. Recently, methods have been developed for removing protein from whey by ultrafiltration. These deproteined whey solutions are ideal sources for the isolation of lactose. After the minerals are removed by ion exchange chromatography, the lactose is then recovered by crystallization from a concentrated solution, followed by centrifugation and drying.

The most common form of lactose is the α-monohydrate, produced from supersaturated solution below 93.5°C. The β-anhydride form is also available but in lesser amounts. It is produced by crystallization above 93.5°C. The α-anhydride form may be produced by drying the hydrate under vacuum at 65°C. When placed into solution, all forms equilibrate to a β-α ratio of 62.25:37.75 at 0°C. The most common uses are in infant formulas, confectionery, and pharmaceuticals. Characteristics of lactose are shown in Table 3-8.

TABLE 3-8. Characteristics of Food/USP Grade Lactose[a]

Component	
Composition	
Lactose, %	≥98.0, db
Protein, %	0.1
Fat, %	0.0
Ash, %	0.1–0.3
Moisture, %	4.0–5.5
pH	4.5–7.5
Color	White to pale yellow
Flavor	Slightly sweet
Microbiology	
Total plate count	<30,000 cfu/g[b]
Coliforms	<10/g
Salmonella	Negative
Listeria	Negative
Coagulase-positive *Staphylococci*	Negative

[a] Data from (7).
[b] Colony-forming units per gram.

TABLE 3-9. Typical Composition of Maple Sugar[a]

Component	Percent
Water	34.0
Sucrose	58–66
Hexoses[b]	0–8
Malic acid	0.09
Citric acid	0.01
Soluble ash	0.3–0.8
Insoluble ash	0.08–0.7

[a] Data from (8).
[b] Including glucose and fructose.

TABLE 3-10. Grades of Maple Syrup[a]

Grade	Color	Color Index[b]
U.S. AA Fancy	Light amber or lighter	0–0.51
U.S. A	Medium amber or lighter	0.51–0.897
U.S. B	Dark amber or lighter	0.897–1.455
U.S. Unclassified	Darker than dark amber	>1.455

[a] Data from (8).
[b] Color index = $A \times 86.3/cm = A_{450}(86.3/bc)$, where A = absorbance, A_{450} = absorbance at 450 nm, b = depth of solution (cm), and c = concentration (g of sucrose per 100 ml).

MAPLE SYRUP AND MAPLE SUGAR

Only two of the maple trees native to North America, the sugar maple and the black maple, are important in maple syrup production; their sap is sweeter than that of other species. The region of syrup production extends from Maine west to Minnesota and from Quebec south to Indiana and West Virginia.

Maple syrup is mainly syrup made by the evaporation of maple sap, which is accomplished in flat, open pans to a concentration of 65.5% solids. Special thermometers relate temperature to concentration; e.g., a 65.5% solution boils at 104°C (219°F).

As it comes from the tree, maple sap contains about 2% solids, of which 97% is sucrose. The rest is organic acids, ash, protein, quebrachitol, polysaccharides, and a trace amount of lignin. However, this composition changes during evaporation, producing some glucose and fructose on inversion at low pH plus small amounts of other saccharides.

Evaporation times and temperatures must be carefully controlled, since color and flavorants increase as the result of processing and it is the trace materials that give the syrup its characteristic maple flavor. One group of flavorants contains ligneous materials from the sap. A second group is formed by caramelization of the sugars. The composition of a typical maple syrup is shown in Table 3-9. Maple syrup is also graded (Table 3-10).

Maple sugar is a solid product made by further evaporation of the syrup to about 92% solids.

FRUIT-DERIVED SWEETENERS

Some relatively new carbohydrate-based sweeteners, used in food formulations, are made from various fruit and/or fruit juice concentrates or from grain-based syrups such as rice syrup. These fruit-derived sweeteners have good *humectancy* and low water activity and are also useful as fat replacers. They can improve shelf life and can function as emulsifiers and antioxidants because of the presence of fruit components such as pectins. They are often used in "natural" foods because they are not labeled as "sugar," which can have a negative connotation in the natural foods market.

Humectancy—The property of retaining moisture.

References

1. Howling, D. 1992. Glucose syrup: Production, properties, and applications. Pages 277-317 in: *Starch Hydrolysis Products*. F. W. Schenck and R. E. Hebeda, eds. VCH Publishers, New York.

2. Day, G. A. 1987. Crystallization of fructose utilizing a mixture of alcohols. U.S. patent 4,643,773.
3. Okada, M., and Nakakuiki, T. 1992. Oligosaccharides: Production, properties, and applications. Pages 335-336 in: *Starch Hydrolysis Products*. F. W. Schenck and R. E. Hebeda, eds. VCH Publishers, New York.
4. Hickenbottom, J. W. 1993. *Malts in Baking Foods*. Tech. Bull. Vol. 5, no. 3. American Institute of Baking, Manhattan, KS.
5. Anonymous. 1988. *Honey: From Nature's Food Industry*. National Honey Board, Longmont, CO.
6. Doner, L. W., and Hicks, K. B. 1982. Lactose and the sugars of honey and maple: Reactions, properties, and analysis. In: *Food Carbohydrates*. D. R. Lineback, ed. AVI Publishing Co., Westport, CT.
7. American Dairy Products Institute. 1992. *Ingredient Description Brochure*. The Institute, Chicago.
8. White, J. W., and Underwood, J. C. 1974. Maple sugar and honey. In: *Symposium: Sweeteners*. G. E. Inglett, ed. AVI Publishing Co., Westport, CT.

Analytical Tests for Sweeteners

Measurement of the physical, chemical, and microbiological properties of sweeteners is important for several reasons. It is vital to the producer to ensure the production of a safe, consistent, high-quality product and equally important to the user, who must be able to identify the right ingredient for a specific product. Both producer and user must be concerned about possible changes in quality during storage and transportation and how those changes might affect functionality.

To assure consistency and quality, standard methods have been established to evaluate sweeteners. For corn-based sweeteners there are the *Standard Analytical Methods* and *Critical Data Tables* of the Corn Refiners Association (1,2). Sugar companies have their own individual procedures, which have essentially become standardized over the years. Generally these procedures are based on methodologies collaboratively tested and approved by scientific organizations such as AACC, American Oil Chemists Society, and AOAC International. Standard microbiological tests are also available in the *Bacteriological Analytical Manual* (3) and in the *Compendium of Methods for the Microbiological Examination of Foods* (4).

Physical Tests

Physical tests that are important for sweeteners are particle size, clarity or color, and viscosity.

PARTICLE SIZE

Particle size, or particle size distribution, of dry sweeteners can affect the sweetener's functionality in many applications. If the particle size is too large, it will give too much grainy or gritty appearance, texture, or mouthfeel to a product, or it will dissolve in water too slowly. If the particle size is too small, the sweetener will cake and not flow properly, or it will create dusting problems during packaging. The size of particles can usually be controlled during the manufacture of a sweetener by adjusting the rate of crystallization, the extent of milling or grinding, or the screen size through which it is passed.

Particle size is frequently measured using U.S. standard screens combined with a standard tapping or shaking procedure. Instrumental techniques are also employed, primarily with finely granulated products. These methods employ instruments such as the Fisher sub-sieve sizer or techniques such as image analysis or laser diffraction (light scattering).

CLARITY OR COLOR

The clarity and/or color of a sweetener in solution can be very important in food applications. For instance, particular colors are associated with particular grades of maple syrup. Color that is too dark suggests overprocessing and possibly off-flavors as well. Cloudiness in a corn syrup may suggest improper starch conversion or crystallization of dextrose, both indicating a substandard product. The clarity or color of a sweetener can be measured to monitor quality, as well as to measure changes during storage.

Clarity (or cloudiness) and color are usually measured spectrophotometrically as absorbance or transmission at a solids level and wavelength specific to that property. Color can also be visually compared to a standard, using a combination of standard Lovibond red and yellow colored glasses.

Honey color is measured using a Pfund color meter. Its color ranges from nearly colorless (water white) to dark brown (dark amber). The color designations, color range on the Pfund scale, and absorbances are shown in Table 4-1. While color is often specified for honey, it is not a quality factor in the U.S. Department of Agriculture grading system.

TABLE 4-1. Color Designations for Honey[a]

USDA Color Designation	Pfund Scale (mm)	Absorbance[b]
Water white	<8	0.0945
Extra white	9–17	0.189
White	18–34	0.378
Extra light amber	35–50	0.595
Light amber	51–85	1.389
Amber	86–114	3.008
Dark amber	>114	...

[a] Courtesy National Honey Board.
[b] Absorbance = log10 (100/percent transmittance), 560 mm for 3.15 cm thickness of caramel-glycerin solutions measured against an equal thickness of glycerin.

VISCOSITY

The thickness or flowability of a sweetener in solution (or the lack of these characteristics) is an important consideration during processing and for final products. If the viscosity is too high, the sweetener may be difficult or impossible to pump through processing equipment. If it is too low, the sweetener may not provide the proper texture or mouthfeel desired in the final food product. Viscosity tends to be more critical with sweeteners like corn syrups or malt syrups, in which viscosity decreases as DE increases.

Viscosity is frequently measured at a specific solids level and temperature using a viscometer, such as the Brookfield viscometer. Viscosity is highly dependent on the solids level of a sweetener, and changes in viscosity may reflect what's happening in the process as evaporation or dilution changes the solids level.

Chemical Tests

Chemical tests for sweeteners include moisture or water content, solids content, dextrose equivalent, saccharide distribution, pH, ash content, and sulfur dioxide level.

MOISTURE OR WATER CONTENT

Moisture or water content of a carbohydrate-based sweetener is an important chemical property. Carbohydrates have an abundance of hydroxyl groups on each molecule, usually three or more per sugar unit. These groups form hydrogen bonds with water molecules, making it difficult in many cases to remove the last traces of moisture from the sweetener during manufacture.

In some cases, as with dextrose, the compound is usually produced as the monohydrate, because that form is the easiest and least costly to manufacture. The various carbohydrate syrups (corn, malt, rice, etc.) are usually sold as liquids with high solids (70–80%, depending on the product). Here water content is very critical: too much water makes the product unstable from a microbiological standpoint; too little water causes the simple sugars to crystallize or the viscosity to become too high to handle.

Moisture can be measured using the Karl Fischer method. The moisture content of solid sweeteners can also be determined by weight loss using a forced-air or vacuum oven for a standard time and temperature (e.g., 2 hr at 70°C at 50 mm of Hg). For corn syrups, oven temperature is generally lower for syrups with high DE. Rapid methods of moisture determination for quality control or production control have been developed using in-line refractometers and near-infrared instrumentation.

SOLIDS CONTENT

Solids content can be determined by a number of methods that are interrelated and therefore can be converted from one scale to another. Appendix C shows interrelationships between three methods. These measurements have common denominators such as *refractive index* or *specific gravity*. Standardized tables have been developed for sucrose, dextrose, fructose, honey, and the various corn, malt, and maple syrups.

Refractive index. A solution of simple sugars in water refracts the light that passes through it. The degree of refraction of the light is called the refractive index. Because this measurement depends upon light, the wavelength of the light used as well as the temperature at which it is taken affect the measurement. Therefore, when reporting the refractive index, the temperature and wavelength must also be given. Tables available through the Corn Refiners Association provide percent solids based on refractive index at specific temperatures for specific corn-based sweeteners.

Specific gravity. The specific gravity of a solution is determined by weighing a specified volume of the solution and the same volume of pure water. Both weights are taken at a standard temperature, usually room temperature or 25°C. The ratio of the mass of the solution to the mass of the same volume of pure water is a unitless number called the specific gravity of the solution.

Refractive index—A physical property of a substance that relates to how light is refracted from the material. Usually used to indirectly measure some other property, such as soluble solids (i.e., the total sugars in solution).

Specific gravity—Ratio of the density of a sample solution and the density of water at the same temperature.

Brix. The Brix measurement is commonly used to determine a specific solids level, namely, the percent sucrose in a solution. The units of measurement are called *degrees Brix*. A 65% sucrose-in-water solution, on a weight/weight basis, would give a measurement of 65 degrees Brix, also written as 65° Brix. The measurement can be made by two different methods, one using a hydrometer and the other a refractometer. Both devices need to be calibrated for measuring degrees Brix. It is important to remember that these measurements relate the specific gravity of a solution to an equivalent concentration of pure sucrose. When measuring solutions with other sugars present (such as glucose, fructose, or honey), the designation "percent solids" should be used. High-fructose corn syrup is often used as a replacement sugar for sucrose in the beverage industry. As a result, special charts that give correction factors for hydrometer or refractometer readings for high-fructose corn syrup are also available.

Baumé. This measurement is based on specific gravity. A hydrometer is used, and the measurement is reported in *degrees Baumé* (°Bé). Because solutions may be too viscous at room temperature, the determination is often made at a higher temperature (e.g., 60°C [140°F]). Tables are available through the Corn Refiners Association that convert Baumé readings into percent dry substances or specific gravity for corn-based sweeteners with given DE and ash contents.

DEXTROSE EQUIVALENT

As explained in Chapter 3, dextrose equivalent (DE) is a measure of the extent of starch hydrolysis for corn-based sweeteners (and malt and rice syrups as well). DE is determined by measuring the amount of reducing sugars in a sample relative to the amount in dextrose. As starch is hydrolyzed, more and more dextrose and other simple sugars are produced, and the amount of reducing sugars and the DE increase. This test is especially important for following the course of hydrolysis and for characterizing in-process and finished-product samples. DE has become the standard notation for specific syrups, syrup solids, and maltodextrins.

The standard DE method involves treating a reducing sugar solution with a standard copper reagent and measuring the amount of cupric ion (Cu^{+2}) remaining after reaction; the more Cu^{+2} that is reduced to Cu^+, the higher the DE. Rapid methods have been developed for quality control and process control using freezing point depression techniques.

SACCHARIDE DISTRIBUTION

Instrumental methods for measuring the presence and/or concentration of specific sugars have been developed based on chromatography, particularly high-pressure liquid chromatography (HPLC). These techniques are much more definitive than DE in testing for a specific sugar or in characterizing a mixture of sugars. Standardized

Degrees Brix (° Brix)—A measure of the density or concentration of a sugar solution. The degrees Brix equal the weight percent of sucrose in the solution.

Degrees Baumé (° Bé)—An arbitrary scale of specific gravities of liquids or solutions.

procedures have been developed with chromatographic columns that can separate, identify, and quantify sugars.

A typical HPLC curve of a hydrolyzed corn syrup is depicted in Figure 4-1, showing the distribution of sugar peaks from glucose to the polysaccharide with a degree of polymerization of 22.

The HPLC procedure can also be used with a differential refractometer to distinguish between disaccharides in a mixture, e.g., between sucrose, maltose, and lactose. It can also determine the amount of inversion in a sucrose solution, measuring peaks for sucrose, glucose, and fructose.

pH

The pH, i.e., the acidity or basicity, of a sweetener is important because it can affect the pH of the food system and therefore its stability. Like most food ingredients, sweeteners (in solution at 10–50% solids) have a pH around neutrality (7.0). Many sugars and syrups are more stable at pH 4.0–6.0. The pH of honey is 3.9.

ASH

The ash, or mineral, content of a sweetener is important because it can affect the flavor or appearance of a food product. Food or pharmaceutical manufacturers usually require low levels of residual ash, generally 0.5% or below. Ash is measured as the residue after a sample of sweetener is ignited in a muffle furnace (e.g., at 525°C for 2 hr). Rapid methods for in-process control usually involve a conductivity measurement.

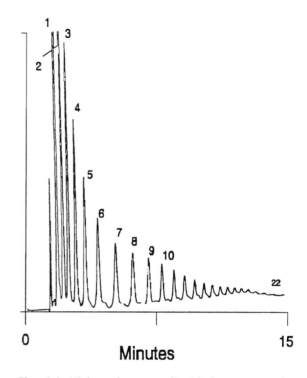

Fig. 4-1. High-performance liquid chromatographic curve of a hydrolyzed corn syrup. Peak 1 is glucose, peak 2 maltose, peak 3 maltotriose, and peak 4 maltotetraose. Peaks 5–22 are oligosaccharides and polysaccharides from degree of polymerization (DP) 5 through DP 22. (Courtesy of Dionex Corporation)

SULFUR DIOXIDE

The sulfur dioxide (SO_2) content of sweeteners, usually reported in parts per million, is especially important because of the allergic reactions and respiratory problems associated with it. Sulfur dioxide is present in some refined sweeteners as the result of their being used as a processing aid (in corn wet milling) or a stabilizer. For determination of sulfur dioxide, a sweetener sample is treated with sodium hydroxide to release sulfur dioxide. The sample is then acidified, and the sulfurous acid is titrated with iodine solution.

Microbiological Tests

Most sweeteners are produced from naturally grown biological products (grain, sugar cane, milk, etc.). Certain types of microorganisms are always present in sweeteners to some extent. They may be present in the raw materials, in the air, in the process water, and potentially on the equipment used. In addition, process conditions are usually conducive to their growth and multiplication. Consequently, levels of microorganisms must be measured and controlled at more than one site to ensure safe food ingredients. Standard microbiological methods are available from several sources.

With most microbiological tests, a solution of the sweetener is streaked onto a plate containing a growth medium specific for the organism being tested. The plate is covered and placed into an incubator for a specific time and at a specific temperature. Finally, the colonies on the plate are counted, and the number of organisms per gram is calculated using the appropriate conversion factor.

The standard plate count, also called total plate count, is a test that measures the total bacteria per gram of sample (or bacteria per milliliter of liquid sample). The same method can be used for measuring yeasts and molds. These microorganisms are sufficiently different that a skilled technician can detect and count colonies of both bacteria and yeasts or molds on the same biological plate. Typical values in sweetener specifications are shown in Table 4-2.

Coliforms, of which *Escherichia coli* is one variety, are pathogenic bacteria that are quantified by the standard plate count. Tests for coliforms should be <10/g. *Salmonella* is a bacteria that causes food poisoning and therefore must not be present in sweeteners.

TABLE 4-2. Typical Microbiological Specifications for Selected Sweeteners

Determination	Corn Syrup[a]	Dextrose	High-Fructose Corn Syrup
Total plate count	<50/g	<5,000/g	200/10 g dse[b]
Mold count	<10/g	⋯	10/10 g dse
Yeast count	<20/g	⋯	10/10 g dse
Osmophilic mold	<20/g	<200/g	⋯
Osmophilic yeast	<10/g	<200/g	⋯

[a] 43 DE.
[b] Dry sugar equivalent. Specifications of the Society of Soft Drink Technologists.

References

1. Corn Refiners Association. *Standard Analytical Methods of the Member Companies of the Corn Refiners Association, Inc.*, 6th ed. The Association, Washington, DC.
2. Corn Refiners Association. *Critical Data Tables*, 3rd ed. The Association, Washington, DC.
3. AOAC International. 1995. *FDA Bacteriological Analytical Manual*, 8th ed. AOAC International, Washington, DC.
4. Technical Committee on Microbiological Methods for Foods. 1992. *Compendium of Methods for the Microbiological Examination of Food*, 3rd ed. C. Vanderzant and D. F. Splittstoesser, Eds. American Public Health Association, Washington, DC.

Chemical and Functional Properties

The chemical properties of a carbohydrate-based sweetener, as well as its functional properties, help determine how it is used to produce the desired effects in food systems.

Chemical Properties

COLLIGATIVE PROPERTIES

Colligative properties are chemical properties that depend on the concentration of the solute in a given solution. They do not depend on what type of molecule is in the solute. For carbohydrate sweeteners, as for other solutes, these properties include freezing point depression, boiling point elevation, osmotic pressure increase, and vapor pressure lowering. Although these properties are valid only when the solute is in a dilute solution, this discussion will assume that the laws for colligative properties hold true for food systems in general.

Because the colligative properties depend only on the number of molecules present, not the type of molecule, sweeteners with different molecular weights (and thus different numbers of molecules on a per-weight basis) have different effects. One molecule of glucose in a system and one molecule of sucrose in another system would each affect a given colligative property in the same manner. However, if the sugars in the two systems were of equal weight (e.g., 5 g of glucose and 5 g of sucrose), their effects on the given colligative property would be different. This is because there are more glucose molecules in 5 g of glucose than sucrose molecules in 5 g of sucrose. Glucose is a monosaccharide and has a lower molecular weight than the disaccharide sucrose.

This fact is important when formulating foods and comparing how different carbohydrate-based sweeteners will affect a colligative property of a food system. For any colligative property, there are clear relationships between sweeteners with increasing saccharide units. Table 5-1 shows the trends for the chemical and functional properties of sweeteners, covering the range from low to high dextrose equivalent (DE). The next sections discuss some of the properties listed in the table. For ease of comparison, it is assumed that the sweeteners are present in the system on an equal-weight basis. However, this

TABLE 5-1. Effect of Dextrose Equivalent Level on Sweetener Properties[a,b]

Property	Dextrose Equivalent					
	0 (Starch)	20	40	60	80	100 (Dextrose)
Sweetness						→ (increasing to the right)
Bodying agent	← (increasing to the left)					
Viscosity	← (increasing to the left)					
Browning reaction						→ (increasing to the right)
Cohesiveness	← (increasing to the left)					
Foam stabilization	← (increasing to the left)					
Freezing point depression						→ (increasing to the right)
Hygroscopicity						→ (increasing to the right)
Humectancy						→ (increasing to the right)
Prevention of ice crystals	← (increasing to the left)					
Solubility						→ (increasing to the right)
Osmotic pressure						→ (increasing to the right)
Fermentability						→ (increasing to the right)

[a] Adapted from (1). Used by permission of the Corn Refiners Association.
[b] Direction of the arrows indicates increase in the property.

may not be true in formulating because of differences in the sweetening powers of the sweeteners.

Freezing point depression. Sweeteners lower the freezing point of products to which they are added. In general, frozen products containing sweeteners are less solid at a given temperature than the same products without sweeteners. The opposite also holds true. That is, the removal of sweeteners causes the freezing point to increase, and frozen products in general become more solid. This property is used to control or prevent the formation of ice crystals in products such as frozen desserts. When the freezing point is lower, it is more difficult for ice crystals to form.

Sweeteners with higher DE have a greater ability to lower the freezing point of the food system than sweeteners with low DE. However, lower-DE (higher molecular weight) materials must be present in frozen products, such as ice cream, to help prevent ice crystal formation and control the size of ice crystals.

Boiling point elevation. Adding sweeteners increases the boiling point of a food system. This factor is important in the manufacturing of confectionery products such as hard candies. Boiling points of candy formulas are often very high (300°F, 149°C) and can change when different sweeteners are added or substituted for one another. Higher-DE sweeteners have greater ability to raise the boiling point than low-DE ones.

Osmotic pressure increase. Adding sugars to food systems increases the osmotic pressure. In fruit and vegetable processing, as the osmotic pressure increases, cell damage decreases and cell preservation results. This principle is used in preserving pickles, relishes, and maraschino cherries. High-fructose corn syrups also increase the osmotic pressure in foods. In formulating enteral fluids, low-DE maltodextrins usually cause fewer diarrheal problems in patients than when higher-DE maltodextrins are used. Osmotic pressure increases as DE increases.

Vapor pressure lowering. The addition of a sweetener to a food system decreases the vapor pressure of that system. As DE increases, the ability to lower vapor pressure increases. This property may be important in the development of foods where aromas are important, such as flavored hot beverages.

DIELECTRIC PROPERTIES

The dielectric properties of sweeteners are such that they enhance the heating of food in a microwave oven. The microwave oven is basically a capacitor with a high-frequency pole charge on its two plates. The food is placed between the plates and takes the role of the capacitor's dielectricum. The high-frequency pole charge interacts with the simple sugar-water binary systems common in food products. It reorients the asymmetrically configured molecules of water and sugar at a very high frequency level. As a consequence, high-intensity friction is created, which warms the product containing the water and simple sugar molecules.

SOLUBILITY AND CRYSTALLIZATION

The solubility characteristics of a sweetener and its crystallization behavior go hand in hand. Sweeteners exist in either a solubilized or crystalline state, depending on the temperature of the system and both the type and concentration of the sweetener. Most carbohydrate-based sweeteners have relatively high solubilities, which influence the scope and nature of their use.

Solubility curves for sweeteners are helpful to product developers. The solubilities of specific sugars are plotted against temperature (Fig. 5-1). Generally, the solubility of the sugar increases as the temperature increases. Since food systems are much more complex than simple solutions and contain many different molecules that may influence solubility, these diagrams should be used only as indicators for sweeteners in food systems. Although the solubility curves depend upon the type of sweetener molecule, trends are evident. In general, solubility increases as the DE increases, with dextrose being the most soluble. With sweeteners of 15–20 DE, solids levels of up to 70–75% can be obtained, while with products of 5–10 DE, the maximum is 60–65% solids. Corn syrups are normally sold and stored as 80% solids, and the more common simple sugars are soluble in the range of 100–400 g/100 g of water. Maltodextrins and corn syrups are quite soluble, but viscosity buildup prevents very high solids dispersions in the lower-DE products. Lactose, although it is a disaccharide, is not a very soluble sugar.

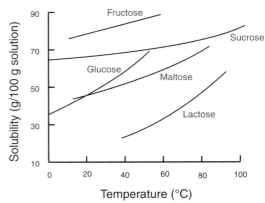

Fig. 5-1. Solubility curves for selected sugars in water. Adapted from (2).

Functional Properties

SWEETNESS AND FLAVOR

The sweetness imparted by a given sweetener in a food product depends on many factors: type of sweetener, temperature of the system, state of the sweetener, and what other molecules are present in the system. In choosing a sweetener in the development process, relative sweetness, character/quality, and synergistic effects must be considered. Sweetness as a property is further discussed in Chapter 2.

Relative sweetness. No two sweeteners have the same sweetness profile and taste spectrum. However, they are often assigned relative sweetness scores. Although other references can be used, the usual standard for relative sweetness comparisons is sucrose, which is taken to be either unity or unity times 100. Table 5-2 shows the relative sweetness scores for selected sweeteners.

Reported values for relative sweetness scores can be expected to vary considerably because they depend on the taste panel method, solution concentration, and temperature of analysis. For those values determined with solutions, the calculation for relative sweetness is made on a weight-per-volume basis. This calculation has practical application in the food industry, but it can be misleading for other applications. For example, on a weight basis, fructose is the sweetest sugar, with a relative score as high as 1.7–1.8, but on a molecular (molar equivalent or equimolar) basis, sucrose is by far the sweetest; fructose is about 0.95. Therefore, the conditions for relative sweetness comparison must be clearly defined.

Temperature also affects relative sweetness. Relative intensities are often determined for sweeteners and plotted versus temperature. Typical relative intensity curves are shown in Figure 5-2. The relative sweetness of fructose drops markedly after it has been dissolved in water and allowed to stand for a short time. Also, when the temperature of a fructose solution increases from 5 to 60°C, its relative sweetness decreases by about 50%. Thus, fructose possess its highest relative value at cold temperatures and loses most of its sweetness at high temperatures. The crystalline form of fructose is β-D-fructopyranose, but in solution at varying temperatures, an equilibrium is established between several other structural forms of fructose. Although one of them has a structure similar to that of sucrose, none of these other forms is particularly sweet. This, however, is not the case for glucose. Changing temperatures have little effect on the balance between the different glucose structural forms. The relative sweetness of glucose solutions, therefore, is not affected by changes in temperature.

Lactose has several crystalline forms, none of which is very sweet. The most common form is α-D-lactose monohydrate, which is only sparingly soluble in water. β-D-Lactose has greater solubility. Because of its properties, lactose is used primarily as a bulking agent rather than a sweetener.

TABLE 5-2. Relative Sweetness Values of Selected Sugars

Type of Sugar	Sweetness
Fructose	1.1–1.7
Sucrose	1.00
Glucose	0.5–0.8
Maltose	0.3–0.6
Galactose	0.3–0.6
Lactose	0.2–0.6

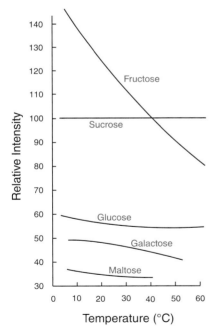

Fig. 5-2. Sweetness intensity curves for selected sweeteners. Sucrose taste intensity = 100 at all temperatures. Adapted from (2).

Flavor quality. The character or quality of a sweetener is another factor to be considered in comparing sweeteners. Taste quality is relative, best judged in the food product being considered. The off-notes, lag times, and sweetness durations noted in product literature may not be relevant to the food system under consideration. Bitter notes or extended duration may be masked by the other ingredients and flavors. Product literature usually cites a time-intensity profile (Fig. 5-3), which identifies relative sweetness intensity over a period of time.

Crystalline glucose has a relatively clean taste spectrum. However, at high concentrations, glucose also appears to have an additional taste that has been described as burning or bitter. Possibly, this is due to the residual acid used commercially in the final conversion of starch dextrins to glucose. Processing technique also affects the taste of sucrose. At one time, the difference in ash composition between beet sugar and cane sugar led to the assignment of a relative sweetness score of 97.9 to beet sugar and 100 to cane sugar, presumably because of the taste impurity resulting from the presence of potassium chloride. In this case, the so-called relative sweetness scores of 97.9 versus 100 are not integrated relative profile scores, but instead are integrated relative taste spectrum scores that affect the sweetness character.

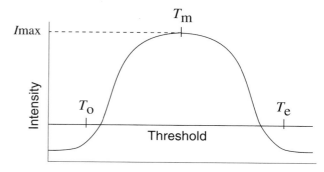

Fig. 5-3. Generalized time-intensity profile for sweeteners. T_o = time of sweetness onset, T_m = time of maximum intensity, T_e = time of extinction, I_{max} = maximum intensity. $T_o - T_e$ = duration of sweetness.

Synergistic effects. A synergistic effect occurs when two sweeteners are combined and their sweetness is greater than would be predicted using the sweetness levels of the two sweeteners individually. For example, fructose is known to provide a synergistic effect when used with other sweeteners.

High-fructose corn syrups have a very bland, sweet flavor and do not mask other flavors in food systems. Instead they tend to enhance many flavors, particularly when used in beverages. This is also true of crystalline fructose. Perception of sweetness is faster with fructose than with glucose or sucrose. The fast perception enhances other flavors, which tend to develop more slowly and can be masked by other sweeteners.

Glucose solutions of 5–10% are approximately one half to two thirds as sweet as sucrose, but over the range of 1–20% glucose in water, the relative sweetness increases by about 50%. This phenomenon is known as self-synergism.

CONTROL OF WATER IN SYSTEMS

Carbohydrate-based sweeteners can influence the state of water in food systems. This control affects many aspects of the final product, including microbial spoilage, or shelf life, and texture. The *water activity* (a_w) of a food is defined as:

$$a_w = p/p_o = ERH/100$$

where p = partial pressure of the food at a given temperature (T), p_o = partial pressure of pure water at T, and ERH = equilibrium relative humidity. The lower the a_w, the more the water is tied up in the system. Because of their ability to form hydrogen bonds, carbohydrate-based sweeteners essentially bind water and thus lower the a_w of a food. This bound water cannot be readily utilized by microorganisms for growth. Therefore, depending on the whole food system, sweeteners can play an important role in food preservation.

A *humectant* is a material that is able to attract water to itself. This ability results from its hygroscopicity, which is the ability to absorb and retain moisture. Many sweeteners are used as humectants in formulations due to their hygroscopic nature. Both these properties increase with increasing DE. As the number of hydroxyl groups (-OH) per sweetener molecule increases, the amount of hydrogen bonding increases and therefore water absorption increases. These factors are important in determining appropriate storage conditions. For example, fructose is a very hygroscopic sugar as compared to sucrose. If both were stored in a humid environment, the fructose would cake and could even harden to brick form, whereas the sucrose might still remain free-flowing.

VISCOSITY

Viscosity is related to the ability of a material to flow. Giving body, or mouthfeel, to a food product is done by providing viscosity and the capacity to suspend particulates. As the molecular weight, or number of saccharide units, increases, the ability to increase viscosity also increases. The lower-DE corn syrups and maltodextrins have higher viscosity and function as bodying agents in canned fruit, caramel candies, and related applications.

Starches are used to increase the viscosity of many food systems because of their ability to gelatinize in the presence of heat and water. In general, the presence of sweeteners causes starches to gelatinize at higher temperatures. This temperature increase is much less when the sweetener is fructose than when it is glucose or sucrose. In other words, a mixture of starch and fructose in water can gelatinize at a lower temperature or to a greater extent at the same temperature than can a mixture of starch and other sugars. Many sugars also inhibit viscosity development in starch. Fructose does not inhibit it to the same degree as sucrose, indicating that starch levels could actually be reduced in some food systems when fructose is used in place of sucrose.

FOAM STABILIZATION

Sweeteners can stabilize foams by increasing the viscosity of the matrix that surrounds the air cells. Also, when the matrix is given

Water activity—The measurement of the degree to which water is bound in a system. The water activity of foods is measured on a scale of 0 (dry) to 1.0 (moist).

Humectant—A material that is able to attract water to itself.

more structural support, the foam is more stable. In this manner, the lower-DE or higher molecular weight sweeteners help to stabilize foam systems. Lower-DE corn syrups or corn syrup solids are the preferred sweeteners for use in stabilizing the foams in dairy whipped toppings, usually in combination with other sweeteners.

COHESIVENESS

Cohesiveness is a measure of how well a material stays together. Sweeteners are often used to hold various components together as one product. As sweeteners with decreasing DE are used, the cohesive properties increase. The lower-DE products are better "glues" than the higher-DE products because they have more large molecules that can associate with molecules of other ingredients as well as with other carbohydrate molecules. In other words, there are many more sites and opportunities for hydrogen bonding. This property is important in holding together products like granola bars.

References

1. Hoover, W. J. 1963. The technological development and increased usage of glucose sirups in the U.S. (Paper presented.) Corn Refiners Association, Washington, DC.
2. Belitz, H.-D., and Grosch, W. 1987. *Food Chemistry*. (Translated from the 2nd German edition by D. Hadziyev.) Springer-Verlag, New York.

Bakery and Other Grain-Based Products

Carbohydrate-based sweeteners are present in the vast majority of grain-based products found in the marketplace. They are used at concentrations that range from 12% in products such as crackers and pizza dough to 25–30% in cakes and muffins (Table 6-1). In "high-ratio" cakes, the amount of sugar is greater than the amount of flour.

This chapter discusses bakery products, which include hard and soft wheat products, breakfast cereals, and granola products. Frostings, glazes, icings, and fillings are covered here because they are integral components of many grain-based products.

Functions of Sweeteners in Grain-Based Products

Although the most obvious use of sweeteners in bakery products is to provide sweetness, carbohydrate-based sweeteners also perform several other functions. These include formation of the characteristic brown color of bakery products, enhancement of flavor, participation in the formation of the product's structure, providing softness, and giving the product's surface the desired appearance (Fig. 6-1).

SWEETNESS

Sweeteners such as sucrose, dextrose, fructose, and high-fructose corn syrup [HFCS] are used to enhance the sweet flavor of the product. Other sweeteners, e.g., lower dextrose equivalent (DE) corn syrups, may be used to add other functionalities, such as to increase browning or add solids to low-fat foods, while contributing a low level of sweetness to the product. Honey, molasses, and refiners syrups are used to add sweetness and also contribute other unique flavors.

BROWN COLOR FORMATION

Sweeteners may be added to produce or enhance the brown colors associated with crusts on baked products, the crumb of toasted breads, the toasted color of cereal flakes, and the brown or golden crumb of muffins and quick breads. There are three sources for these brown colors. The Maillard browning reaction and caramelization reactions, discussed in Chapter 1, are two of these sources. The third

TABLE 6–1. Concentration Ranges of Sweeteners in Baked Goods[a]

Product	Percent Sweetener Used
Pie dough	1[b]
Crackers	1–2
Pizza dough	1–3[c]
Biscuits	2–3
Yeast-raised doughnuts	2–4
Bread and rolls	2–10[d]
Pastry or Danish dough	5–10[d]
Sweet breads or sweet rolls	7–15
Cake doughnuts	15–20
Cookies	20–25
Muffins	25–30
Cakes	25–30

[a] Data taken in part from private communication from M. Olewnik of the American Institute of Baking.
[b] Usually as dextrose for browning.
[c] The amount used depends on type of crust and level of fermentation desired.
[d] Level indicated or more depending on amount of sweetness required.

Fig. 6-1. Bakery products come in a wide variety of shapes and textures and may include fillings, frostings, and other toppings.

source of brown hues is the pigments in brown sugar, malt syrups, molasses, refiners syrups, and honey.

FLAVOR ENHANCEMENT

The sweet flavor of grain-based products can range from clean, sweet notes to toasted brown or burnt sugar notes. Maillard browning and the caramelization reaction are responsible for many of these flavors. Sweeteners also contribute many backnotes of the flavors in the products. The sulfury notes of molasses or the distinctive floral notes from honey are difficult to duplicate synthetically. Sweeteners also tend to enhance and complement other flavors present. One of the most notable examples is the mellowing and increased sweetness of cinnamon that occurs when sucrose is added. Also, sweeteners cause different sensations on the tongue (e.g., cooling effects from dextrose, a cleaner taste from sucrose) that affect the choice for topical applications.

STRUCTURE

One of the hidden roles of sweeteners results from their ability to affect structure and texture. This is most evident when a manufacturer reformulates a standard product to a sugar-free product. Many sweeteners delay starch gelatinization and increase the temperature at which proteins denature and coagulate, both of which are important in setting the structure of cakes. In various cookie formulations, sucrose creates crispness and snap. In granola bars, sweeteners such as honey and corn syrups give the product cohesiveness, enabling the components to stay together as one piece. Often, ready-to-eat (RTE) cereal pieces keep their crispness in milk because of the hard shell coating of a sweetener glaze.

SOFTNESS

Honey and sweeteners with a high DE have the ability to attract water. This attribute is important to the softness of the crumb in baked goods. Also, in granola formulations, these sweeteners allow the products to break apart easily without becoming too hard to the bite or so brittle that they fall apart.

APPEARANCE

Different appearances can be attributed to the use of specific sweeteners. Cracks on the tops of cookies can appear when sucrose is used; if fructose is the sugar, no cracks appear. The frosted appearance and the shinier glazed look of sugar-coated RTE cereal pieces are also influenced by sweetener choice. Sweetener selection can create the sheen of a bakery glaze and the matte, opaque appearance of a frosting. Topical sugar coatings on doughnuts and cookies are dependent not only on the type of sweetener, but also on its granular size.

Hard Wheat Products

BREAD, ROLLS, AND BAGELS

Carbohydrate-based sweeteners are typical ingredients in the manufacture of breads. Sucrose is the most commonly used (Table 6-2). However, HFCS is comparable to sucrose in sweetness and can completely replace sucrose in bread and rolls. Other characteristic sweeteners such as molasses and honey can add rich flavors and colors to specialty breads and bagels. Table 6-3 shows some typical levels of honey in baked goods. If corn syrup is utilized as the sweetener, the higher-DE (54–62) products are preferred because of higher sweetness.

In addition, sweeteners such as sucrose (and, in some large-scale processes, HFCS) are often added for their contribution to fermentation, brown crust formation, and texture. Dextrose is good for high-speed bread processes because it is more rapidly fermented than most other sweeteners. High-DE corn syrups also have high fermentability.

In the discussion below, sucrose is used to discuss the functions that sweeteners impart. However, other simple carbohydrate-based sweeteners have a similar effect or exhibit the functional properties discussed in Chapter 5.

TABLE 6-2. Concentration of Sweeteners in Typical Baked Goods

	Percent Sweetener	
Product	Sucrose	Other
Hard wheat		
Pie dough	1.5	
Pizza dough	1.5	3.0 maltodextrin[a]
White bread	3.0	
Whole wheat bread	2.2	4.0 molasses
Dinner rolls		4.0 HFCS[b]
Sweet dough	7.6	
Danish sweet dough	8.25	
Cake doughnut	22.0	1.0 dextrose
Soft wheat		
Pancake mix	7.6	6.0 maltodextrin[a]
Cake		
White	23.0	
Chocolate	21.0	
Brownies	18.0	1.5 corn syrup
Cookies		
Sugar	12.5	12.5 powdered sugar
Chocolate chip	11.7	1.6 corn syrup, 5.5 HFCS,[b] 15.6 brown sugar
Gingersnap		8.0 molasses, 24.0 brown sugar
Peanut butter	16.0	5.0 maltodextrin,[a] 16.0 brown sugar

[a] 10-DE maltodextrin.
[b] High-fructose (42%) corn syrup.

Fermentation. In several procedures for the commercial manufacture of breads, sucrose is added to aid in achieving a consistent final product. The flour used in breadmaking is not an adequate source of the simple sugars needed for yeast fermentation. A shortage of these carbohydrates leads to incomplete fermentation within the time parameters of processing, resulting in a failed product. The addition of sucrose ensures that the yeast will have an excess of easily fermentable carbohydrates. Depending on the desired product, the level of sucrose can range from about 1 to over 10% on a flour-weight basis. Table 6-4 shows formulas of some typical breads.

Yeasts preferentially use monosaccharides as a carbohydrate source for metabolism. Once the monosaccharides are utilized, the yeasts enzymatically break down disaccharides and other relatively simple carbohydrates, followed by other complex carbohydrates. The yeasts enzymatically break down sucrose into its component sugars, glucose and fructose. Glucose is more readily used as a nutrient source than fructose; some residual fructose remains after fermentation (1). The by-products of fermentation are ethanol and carbon dioxide gas. The ethanol is vaporized during baking, and the carbon dioxide is entrapped by the dough matrix, causing formation and expansion of gas bubbles. The bubbles cause the bread to rise, and they form the air cells of the bread crumb.

Brown crust formation. The formation of brown pigments depends on two reaction processes: Maillard browning and caramelization. Sucrose is unable to participate in Maillard browning because it is a nonreducing sugar. However, about half of the sucrose in a white bread formulation is fermented by the yeast, and some of the simple sugars resulting from the fermentation participate in Maillard browning. Some of the sucrose that is not fermented undergoes caramelization (see Chapter 1) to form additional brown pigments and the characteristic flavors associated with the brown crusts of bread products. To achieve more browning, other sweeteners can be added. The higher the DE, the more efficient the sweetener in producing color as a result of the Maillard reaction.

Texture. The unfermented sucrose as well as the fructose and glucose generated during fermentation are important to the texture of the final product because they affect the hydration of the protein molecules. During kneading, gliadin and glutenin protein molecules combine to form gluten. The sugars hydrate preferentially, so if a large

TABLE 6-3. Typical Amounts of Honey in Hard and Soft Wheat Products Made with Honey

Product	Honey, %
Pie dough	3.7
Pizza dough, thin crust	1.96
White bread	3.0
Whole wheat bread	1.4
Dinner rolls	4.3
Sweet dough	7.4
Chocolate cake	29.0
Brownies	17.9

[a] Data from National Honey Board.

TABLE 6-4. Typical Bread Formulations (% flour basis)[a]

Ingredients	Bread Type					
	White Pan	White Hearth	Wheat	100% Whole Wheat	Rye	Raisin
Flour						
White	100	100	60		60–80	100
Whole wheat			40	100		
Rye, medium					20–40	
Water	64	55	59	68	64	64
Yeast	3	2	2.5	3	2	4
Yeast food	0.2–0.5	0.375	0.5	0.25		0.5
Shortening	2	2	3	3	2	3
Sweetener	6–8[b]	2	15[c]	8[d]	0–4	8
Salt	0.2–0.5	2	2.25	2.25	2	2.25
Nonfat dry milk	0.5–3			3	0–5[e]	4
Other			2[f]	2[f]		63.04[g]

[a] Adapted from (2).
[b] Solids basis.
[c] Dextrose, 5%, and molasses, 10%.
[d] Sugar, 6%, and honey, 2%.
[e] Rye sour, commercial.
[f] Vital wheat gluten, 2%.
[g] Raisins, 63%; cinnamon, 0.4%.

amount of sugars is present, the protein hydration necessary for gluten formation is hindered and a very dense, heavy final product results. This can be a defect in the manufacture of white breads and rolls but is a desired outcome in dense wheat or whole-grain products.

Sugars also may extend the shelf life of the final product by helping it to retain moisture. The staling or firmness of bread products results mainly from starch retrogradation, a complex phenomenon involving recrystallization. If the retrogradation is delayed, through the use of emulsifiers, for example, moisture retention becomes a primary factor determining the shelf life of baked products. Sugars or honey bind moisture, which increases crumb softness and moisture, thereby increasing consumer acceptance and extending shelf life.

SWEET DOUGH PRODUCTS

In sweet dough products, the sugar level is much higher than in typical breads and rolls. These products also contain more fat as well as other ingredients such as eggs and milk. They are sweeter and very flavorful because of the amount and types of sweeteners used. Principal sweeteners include sucrose, HFCS, and corn syrups. Brown sugars, molasses, and honey are often the main flavor contributor in these products, making sugar-free baked products the ultimate challenge for product developers.

Fermentation. The addition of carbohydrate-based sweeteners, principally sucrose, is essential to the fermentation process, which is important in products such as yeast-raised doughnuts; cinnamon rolls; and croissants, Danish pastries, and other laminated doughs. Because these products contain more sugar than breads do, yeasts are inhibited and the fermentation time must be increased.

Brown color formation. As in breadmaking, the sugars added to sweet dough formulations undergo Maillard browning and carmelization. Sucrose does not participate in Maillard browning but will undergo carmelization at higher temperatures. In several sweet dough products such as pastries, a golden hue is often preferred over the darker brown colors associated with bread crusts. Baking temperatures are often lower compared to those for bread and are carefully controlled in order to assure the desired golden effect.

Softness. The high sugar level of sweet rolls and yeast-raised doughnuts helps to create a soft and tender crumb. The sugars impede the formation of gluten and therefore yield a product that is not as tough. Fats, especially in laminated dough products, also contribute to the lighter, softer texture. Sugars also cause tenderness due to their ability to hold the moisture contributed by water-containing ingredients such as milk and eggs.

Soft Wheat Products

Creaming—High-speed mixing. In baking, the creaming step mixes fat and sugar.

In many bakery uses, final volume, crumb texture, and cell size are key attributes of an acceptable product. Sucrose plays an important role in achieving these and, with certain products, only partial replacement with other sweeteners is possible (Table 6-2).

The sweetener levels used in products like cakes, muffins, quick breads, cake doughnuts, pancakes, and waffles are often quite high. To achieve moist, tender fine-grained crumb, some cakes are made with high-ratio formulas, in which the weight of the sugar is greater than the weight of the flour. Again, the sweetener most often used is sucrose, especially since most of these products are manufactured and sold as dry mixes. Cakes and cookies produced on a large scale employ a combination of sucrose, corn syrup, HFCS, and corn syrup solids (CSS); a few products also use dextrose. Other sweeteners such as honey (Table 6-3) and refiners syrups are used to impart different flavors.

CAKES

Creaming. The most important use of sugar in cake making is in the *creaming* step. During mixing, tiny air bubbles become entrapped in the fat that surrounds the sugar particles. Carbon dioxide released from the chemical leavening system and steam generated during baking fill these air bubbles, which become the air cells in the final cake. The finer and more numerous the air bubbles, the finer the air cells, or grain, in the final cake crumb.

The crystalline structure of the lipids is important to the air incorporation process. The β′ crystalline form of the triacylglyceride works best. The crystalline form of the sucrose is also important. Because the sugar must be in granular form to ensure proper creaming, use of sucrose in this step is essential. Baker's special sugar (defined in Box 6-1) is known

---

Box 6-1. Special Sugars for Baked Products

Baker's special: a granulation of sucrose developed for its uniform crystal size, which creates uniform cell size in cakes. It is also used for topical applications in cookies and doughnuts. Because its grain size is smaller than that of typical table sugar, it has the ability to dissolve rapidly.

Sanding sugars: sugars used as topical, dusted applications on bakery items. They have a smaller grain size than table sugar yet can still reflect light off the crystalline faces of the grains.

Colored sugars: sugars with colorants added; used to decorate bakery products. Most often, food colorants are added to sanding sugars.

Gel grain: a granulation of sucrose coarser than powdered sugar yet finer than standard sugar. It is very uniform in particle size and therefore is used in dry mix applications so that the sugar remains uniformly distributed in the mix during distribution and handling.

Powdered sugar: the finest granulation of sucrose crystals. Small amounts of starch are added as a free-flow agent to prevent caking problems. Powdered sugar is often used dry as a topping on doughnuts and also in the manufacture of icings, glazes, and frostings, where graininess is a fault and smoothness is important.

Caster sugar—An ultrafine granulated sugar useful for fine-textured cakes and meringues and, because it dissolves easily, for sweetening fruits and iced drinks.

to be the best granulation of sucrose for creaming. In Great Britain, *caster sugar* is also very popular.

Dry cake mixes available in retail stores use emulsifiers to aid in the incorporation of air cells into the final product. For these applications, it is more important that the sugar not segregate from the other ingredients during distribution. A smaller or larger granule size of sucrose may be chosen depending on the particle size of the other ingredients in the mix.

Structure. Sugar plays an important role in setting the final structure in cakes. The presence of sugar increases the temperatures needed to gelatinize starch and to denature proteins. These reactions are responsible for setting the cake's structure.

Texture. Sugars contribute to tenderness of the final crumb. Their ability to bind water contributes to moist crumb. Low-DE syrups are added for humectancy control, and medium-DE syrups are used to control sugar crystallization, maintain moisture balance and emulsification, and improve eating quality. High-DE syrups increase the shelf life and the final volume of the product. CSS can also be used to provide additional viscosity and to improve moisture retention and shelf life in cakes, while dextrose can improve cake volume and eating properties.

COOKIES

Many varieties of cookies are available today. A cookie can be soft and chewy or hard and fracturable. Doughs can be molded into shapes, extruded, wire cut, stamped out, or deposited. Sweetener choice and level are very important in achieving the type of final product desired. Cookies employ a variety of sweeteners, mainly sucrose plus HFCS and corn syrup, with some dextrose, fructose, honey, brown sugar, and molasses in specialty products. Cookie mixes contain mainly sucrose with some HFCS, CSS, and dextrose.

Air incorporation. As in cake making, a granular sweetener, namely sucrose, is essential in the creaming step for cookies. Air is entrapped and, during the baking process, carbon dioxide gas and steam fill these areas to make the product rise.

Structure. Sweeteners affect the setting of the cookie structure as a result of their ability to delay starch gelatinization and aid in the denaturation of proteins, as described for cake making.

Cookie spread. During baking, the granular sweeteners melt and become liquid. This allows the dough to become more fluid and spread out. The amount and types of sweeteners used therefore affect the final cookie spread. However, other ingredients such as fats and eggs also affect cookie spread. Therefore, it is important to experiment with each formula to achieve the desired spread.

BAKERY AND OTHER GRAIN-BASED PRODUCTS \ 53

Texture. Upon cooling of the baked product, the sucrose recrystallizes and hardens, giving the cookies fracturability or "snap." As sucrose crystallizes, water is released; this is absorbed by the flour and other ingredients in the cookie formula. Use of high-DE syrups such as HFCS decreases the ability of sucrose to recrystallize and hence produces a chewier, softer product. Honey can also be used to affect texture; large amounts increase softness and small amounts increase crispness. The amount of other ingredients in relation to the amount of sucrose also affects the extent of sucrose recrystallization; other ingredients interrupt the recrystallized matrix and therefore decrease the fracturability of the final product.

Surface cracking. In some products, surface cracking is a desired attribute. Cracking results from the drying out of the surface in combination with sucrose recrystallization. To reduce surface cracking, high-DE syrups or hygroscopic sugars such as fructose should be used.

OTHER PRODUCTS

Muffins and quick breads. Muffins and quick breads have a grain similar to that of cakes. Also, as described for cakes, sweeteners are important in the delay of starch gelatinization and protein coagulation to set the structure. However, because the crumb formation is achieved by chemical leavening rather than by creaming, there are fewer limitations on the choice of sweetener. High-fructose corn syrups, molasses, and honey can all be used in these products as the main sweetener because no creaming takes place and therefore no granular sweeteners are required. The fructose and glucose they contain in high concentrations act as humectants and can therefore provide very moist final products.

Pancakes and waffles. While sweeteners are used in these products, they are usually at a lower level than in cakes and muffins. Their use in setting the structure of the products is the same as in muffins and quick breads.

The brown color of pancakes and waffles results mainly from the carmelization of the sugars present, although some Maillard browning occurs. The high griddle temperature is responsible for the speed at which carmelization browning takes place.

Although sweeteners are responsible for the moist crumb of baked goods, the addition of oil in muffin and quick bread formulations helps to produce a tender product. High-DE syrups also facilitate production of a moist, soft crumb.

Frostings, Icings, and Topical Applications

Frosting just wouldn't be the same without sugar because many of its attributes are provided by sugar. Other sweeteners such as corn syrups and molasses are also used as flavorants, but the most widely used sweetener is sucrose. Other applications include icings, topical dustings, and creme fillings.

FROSTINGS AND FILLINGS

In general, frostings and fillings are prepared by whipping a fat source such as shortening with sugar to create a creamy consistency. Colors, flavors, emulsifiers, or other enhancing ingredients such as cocoa or cream cheese are added to create distinctive frostings that complement what they coat. Sucrose provides not only sweetness but also bulk and volume to a frosting by providing sites for air to be entrapped when the fat is whipped.

The granulation size of the sucrose is of utmost importance. If the crystals are too large, the resulting frosting will be too grainy or sandy to the tongue. Fine crystals produce a smooth, creamy frosting texture. Powdered sugar is the preferred sugar for frosting manufacture. It is available in a variety of granulations such as ultrafine and fine.

ICINGS

Icings differ from frostings in appearance and consistency. Frostings usually have a matte finish and are thicker, while icings can be either glossy or matte and are thinner. Icings are also prepared differently. Usually, an icing is boiled, although there are many varieties of icing formulation. Granular sucrose or other sweeteners such as corn syrups or invert sugar are also used as icing ingredients.

The granulation of the sweetener is not crucial as in frosting production since the crystals are dissolved and then cooled. Upon cooling, the sugars recrystallize. Inhibitory agents such as syrups can be added to create an icing that has a glossy appearance. The syrup acts as a humectant and attracts water to the surface of the icing.

TOPICAL APPLICATIONS

Sweeteners provide visual enhancement when they are applied to the surface as a dusting on products such as doughnuts or sugar cookies. The most common sweetener used for topical application is dextrose, which has a slight cooling effect on the tongue due to its negative heat of dissolution. Fructose is not a good topical sugar because of its hygroscopicity. Over time, fructose dissolves and cannot be seen on the product.

Ready-to-Eat Cereals and Granola Products

Sweetened RTE cereals are very common in the marketplace. In fact, unsweetened cereals are far fewer in number. Sweeteners such as malt syrup, corn syrups, sucrose, and dextrose are commonly used in formulations (Table 6-5). Cereals are produced either by batch

TABLE 6-5. Sweetener Content of Ready-to-Eat Breakfast Cereals[a]

Cereal	Percent Sweetener		
	Total Sweetener	Sucrose	Other
Corn flakes	6.8	2.6	4.2 HFCS[b,c]
Oat	2.8	2.8	
Rice			
Crispy	8.8	7.6	1.2 HFCS[c]
Puffed	0.1	0.1	
Wheat			
Flakes	9.9	8.2	1.7 brown sugar and honey
Puffed, plain	1.4	0.6	0.8 brown sugar and honey
Shredded	0.4	0.4	
Bran flakes	12.1	9.3	2.8 HFCS[c]
Wheat and malted barley			
Flakes	12.4	6.5	5.9 malt syrup
Nuggets	9.1		9.1 malt syrup
Wheat bran	16.4	13.3	3.1 corn syrup and molasses
Corn flakes, sugar-coated	39.6	37.6	2.0 HFCS[c]
Rice, crispy, sugar-coated	39.0	37.7	1.3 HFCS[c]
Wheat			
Puffed, sugar-coated	45.1	38.0	7.1 brown sugar and honey
Shredded and frosted	24.6	24.6	
Bran flakes with raisins	26.6	10.1	16.5 HFCS[d]
Granola, with raisins	27.4	17.0	10.4 brown sugar and honey
Wheat, puffed, sugar- and honey-coated	57.4	44.2	13.2 brown sugar and honey

[a] Data, in part, from (3).
[b] High-fructose corn syrup.
[c] Sometimes a mixture of corn and malt syrups is used instead of HFCS.
[d] Sometimes a mixture of corn syrup and honey is used instead of HFCS.

processes, such as the batch cooker and pelletizer used in producing products like corn flakes, or continuous processes, such as the twin-screw extruders used in making puffed cereal pieces. Sweeteners can be added at various points in these processes: in the mix, during the cooking process, or as a topical application after the cereal piece has been dried or toasted. The type of sweetener chosen depends on the various attributes desired.

PROCESSING AID

Sweeteners act as a lubricant during extrusion or pelletizing. They compete for the water present in or added to the formula. Hydration of the sweetener inhibits hydration of the starch and delays or decreases the swelling and gelatinization processes.

FLAVOR

Sweeteners such as malt syrup, honey, and refiner's syrups help to add the characteristic flavors associated with products such as corn flakes and flaked wheat products. Syrups such as refiner's syrups are more popular than brown sugar because of their ease of use in processing. These liquid sweeteners can be pumped into a batch cooker or added on-line to a screw-type extruder, whereas addition of bagged, granular sugar products is more labor intensive.

Sucrose is often added to the formula to provide sweetness, but heat from the cooking process causes the Maillard browning reaction to occur, adding a variety of complex flavors to the products. Production of these flavors can be controlled through sweetener selection. Higher-DE corn syrups react more readily than sucrose. If less browning with low sweetness is desired, a maltodextrin may be more appropriate. In addition, the longer the cooking process takes, the more flavors produced. Batch cooking processes, although time intensive, are used in the industry because they create these complex flavors. The twin-screw extruder is commonly used in cereal production because of its speed, productivity, and versatility, but its ability to produce these complex Maillard browning flavors is limited by the reduced time of cooking. Adaptations to the extruder as well as formulation changes such as increasing the amount of reducing sugars can help.

Caramelization during the toasting of flaked products is also essential to the flavor of the final products. Once the cereal dough is extruded or pelletized, it can be dried, flaked, and then toasted. During toasting, the surfaces of individual cereal pieces can be surrounded by air temperatures of 204.4–232.2 °C (400–450°F). Sucrose then undergoes the caramelization reaction, producing brown colors and toasted flavors.

COATINGS

Sweeteners are also used in coating cereal pieces, which helps to increase the crunchiness of the cereal as well as providing sweetness.

Most sugar coatings are applied as an aqueous slurry that is heated to boiling and then sprayed onto the cereal pieces while they tumble in a coating drum (Fig. 6-2). The coated pieces are then dried, cooled, and packaged. Coatings range in appearance from a white, opaque frosted look to a clear, shiny glaze, depending on the sweetener selected. A frosted appearance results from the recrystallization of sucrose; the crystals refract the light, causing the white, opaque appearance. A glazed appearance is achieved when the recrystallization of sucrose is inhibited by making the coating a thick, viscous syrup. Corn syrups, honey, or refiner's syrup help to achieve this effect. High-DE syrups can also be used, although the final product may remain sticky after drying or become sticky in a humid environment. Although maltodextrins can help to achieve a glazed look, they may also lower the overall perceived sweetness of the cereal piece.

Fig. 6-2. Cereal pieces tumbling in a coating drum. (Courtesy of Spray Dynamics)

ADHESION

Granola cereals and bars depend on adhesive ingredients to help them stay together. High-DE sweeteners such as HFCS and honey are used. Maltodextrins are also helpful in formulations, especially when the product may become too sweet if higher-DE syrups are used at an increased level in the formula.

Troubleshooting

The flour-to-sweetener ratio is important in formulating baked products.

A frequent error in working with sweeteners is to neglect the water content of products like HFCS, corn syrup, and dextrose (usually sold as the monohydrate). The moisture content of the sweetener must be checked or calculated and the amount of free water addition must be adjusted accordingly.

The hygroscopicity of sweeteners affects the moisture in the products.

BAKED GOODS—GENERAL		
Symptom	**Causes**	**Changes to Make**
Limited shelf life; product dries out and stales rapidly	Inadequate water absorption/retention	Increase sweetener level. Use sweetener with higher DE or more hygroscopicity.
Batter too thin or doughs too sticky	Moisture content too high	Reduce sweetener level or reduce DE of sweetener.

YEAST-RAISED PRODUCTS		
Symptom	**Causes**	**Changes to Make**
Low volume, long proof time, no break and shred, small air cells, or dense crumb grain	Insufficient fermentation and gas production	Increase fermentation activity by increasing time, temperature, or enzymatic activity. Increase sweetener content. Use sweetener with higher DE.
Lack of crust color and flavor development	Insufficient Maillard browning or caramelization	Increase fermentation activity by increasing time, temperature, or enzymatic activity. Increase sweetener content. Use sweetener with higher DE.
Dark or dull color	Excessive Maillard browning or caramelization	Reduce sweetener content. Use sweetener with lower DE. Decrease malt level.
Crust too thick	Excessive fermentation	Reduce sweetener content. Use sweetener with lower DE. Decrease fermentation activity by decreasing time, temperature, or enzymatic activity.
Excessive expansion with large, non-uniform cell structure or open crumb grain	Excessive fermentation and gas production	Reduce sweetener content. Use sweetener with lower DE. Decrease fermentation activity by decreasing time, temperature, or enzymatic activity.
Off flavors or odors	Microbial or yeast growth in syrup products	Check quality of incoming ingredients.
Tough crumb	Overdevelopment of gluten	Increase sweetener content to limit gluten development.
Batter too thick, low volume, sticky or gummy crumb structure, and/or dense grain	Sweetener level too high or level of hygroscopic sweeteners too high to sucrose.	Decrease sweetener level. Decrease ratio of glucose-containing corn syrups

DOUGHNUTS

Symptom	Causes	Changes to Make
Low volume, dense texture, and/or poor handling	Sweetener level too high or level of hygroscopic sweeteners too high	Decrease sweetener level. Decrease ratio of glucose-containing corn syrups to sucrose. Add dextrose.
Tough texture	Excessive gluten development	Increase sweetener content to limit gluten development.

CAKES

Symptom	Causes	Changes to Make
Low volume	Insufficient air incorporation during creaming step	Increase particle size of sweetener.
Gummy crumb	Sweetener level too high or level of hygroscopic sweeteners too high	Decrease sweetener level. Decrease ratio of glucose-containing corn syrups to sucrose.
Collapsed structure	Starch gelatinization inhibited	Decrease sweetener level.
Tough, dry crumb	Excessive gluten development	Increase sweetener level to limit gluten development. Add corn syrup solids to aid in moisture retention.
Crust color too dark	Excessive Maillard browning or caramelization	Decrease reducing sugar level or increase sucrose level (if possible). Reduce bake time.
Coarse, irregular crumb	Improper creaming	Increase creaming time. Increase sweetener particle size for better air incorporation during creaming.
Low volume or dense crumb for angel food cakes	Lack of foam development or low foam stiffness	Increase particle size of sweetener used.

COOKIES

Symptom	Causes	Changes to Make
No snap, texture too soft	Insufficient sugar recrystallization	Increase sucrose level. Decrease hygroscopic sugars such as brown sugar, fructose, or high-DE syrups. Decrease liquids in formula.
Excessive surface cracking	Too much water absorption by sucrose or absence of reducing sugars	Decrease sucrose level. Increase hygroscopic sugars such as brown sugar, fructose, or high-DE syrups.
Texture too hard	Sugar recrystallized throughout product or low moisture content	Decrease sweetener or sucrose level. Increase hygroscopic sugars such as brown sugar, fructose, or high-DE syrups.
Color too dark	Excessive Maillard browning or caramelization	Decrease reducing sugar level or increase sucrose content. Decrease baking temperature and/or time.

READY-TO-EAT CEREALS		
Symptom	**Causes**	**Changes to Make**
Clumping after coating	Insufficient drying or sweetener solutions too hygroscopic or sticky	Increase drying time. Increase agitation during product coating and/or drying. Decrease hygroscopic sugars such as HFCS, brown sugar, fructose, or high-DE syrups. Increase sucrose-to-hygroscopic sweetener ratio.
Poor bowl life	Insufficient or nonuniform coating	Check process to increase coating application (e.g., increase residence time in coating stage). Select appropriate sweeteners to adjust coating viscosity for more uniform application.
Coating too shiny	Coating is amorphous or glasslike	Increase sucrose-to-hygroscopic sweetener ratio. Increase dryer temperature/time.
Undesired frosted or matte appearance	Excessive sugar crystallization	Decrease sucrose-to-hygroscopic sweetener ratio. Decrease dryer temperature and/or time.

CEREAL-TYPE BARS (GRANOLA BARS)		
Symptom	**Causes**	**Changes to Make**
Texture too hard	Excessive sugar crystallization	Decrease sucrose-to-hygroscopic sweetener ratio.
Texture too soft/chewy	Moisture content too high	Increase sucrose-to-hygroscopic sweetener ratio.
Crumbly cereal bar	Poor binding or cohesive properties of sweeteners	Decrease sucrose-to-hygroscopic sweetener ratio. Apply more sweetener coating.

FROSTINGS AND ICINGS		
Symptom	**Causes**	**Changes to Make**
Gritty, sandy texture	Excessive sugar crystallization or sweetener particle size too large	Decrease particle size. Increase cooking time and/or temperature for boiled icings.
Cracking or formation of a hard shell	Excessive drying	Add hygroscopic sugars such as HFCS, brown sugar, fructose, or high-DE syrups.
Icing does not stick to product or sticks to wrapper or packaging	Moisture too high	Decrease liquid levels. Decrease hygroscopic sugars such as HFCS, brown sugar, fructose, or high-DE syrups. Allow product to completely cool before finishing. Add emulsifier/stabilizer.
Texture too soft	Moisture too high	Decrease liquid levels. Decrease sweetener particle size. Decrease hygroscopic sugars such as HFCS, brown sugar, fructose, or high-DE syrups.

References

1. Ponte, J. G., Jr., and Reed, G. 1982. Bakery foods. In: *Industrial Microbiology*, 4th ed. AVI/Van Nostrand Reinhold, New York.
2. Ponte, J. G., Jr. 1981. Pages 9-26 in: *Variety Breads*. B. S. Miller, Ed. American Association of Cereal Chemists, St. Paul, MN.
3. Matthews, R. H., Pehrsson, P. R., and Farhat-Sabet, M. 1987. *Sugar Content of Selected Foods*. Home Economics Research Report 48. HNIS, USDA, Washington, DC.

Supplemental Reading

Kearsley, M. W., and Dziedzic, S. Z. Use of glucose syrups in the food industry. 1995. In: *Handbook of Starch Hydrolysis Products and Their Derivatives*. Chapman Hall, London.

Confections

Everyone has a favorite kind. For some, it's a fresh, soft, buttery caramel; for others it's the smooth, bitter sensation of dark chocolate. It could even be Grandma's fudge or homemade peanut brittle. There are so many types of confections that it's almost impossible to name them all. But, they all have one thing in common: they are sweetened with carbohydrate-based sweeteners.

The sweetener most abundantly used in confections is, of course, sucrose. Several types of sugars are specially made for confections, including confectioner's sugar, coarse granulated sugar, sanding sugars, powdered sugar, and invert sugar (Box 7-1). However, other sweeteners can be used to aid in the manipulation of such characteristics as

Box 7-1. Special Sugars for Confections

Confectioner's sugar: a very large crystalline sugar, measuring about 3 mm in length and used primarily for its purity and very bright white color. This feature makes it beneficial in the making of fondants and where clear or white products are desired. It is a very hard crystal and is not usually used unless it is dissolved or melted and recrystallized into smaller crystals.

Coarse granulated sugar: sugar crystals slightly smaller than those of confectioner's sugar, often used for colorless or white candies.

Sanding sugars: finer sugars that can be fractured by biting with the teeth. They also tend to reflect light and can add a glittery appearance to a confection. To increase sparkle, these sugars can be washed in alcohol to remove sugar dust from the crystalline faces. They are most often used as a topical application on gelled or gummed candies.

Powdered sugar: a very fine granulation of sucrose, often used for topical dustings and fondant making. Cornstarch is often added to prevent caking.

Invert sugar: a mixture of sucrose, glucose, and fructose. Some of the sucrose present is "inverted" to its components glucose and fructose by treatment with the enzyme invertase. Sucrose also inverts to glucose and fructose in the presence of acid.

texture, graininess, and sweetness. Syrups such as high-fructose corn syrup (HFCS) or high- or low-dextrose equivalent (DE) corn syrups are often employed to retard sucrose crystallization in hard candy formulations or to increase viscosity or chewiness in caramel or taffy. Honey can also achieve these effects; however, some dark honeys contain strong, distinctive flavors that may be too strong for certain confections in which a clean, sweet flavor is desired. Brown sugars can also add flavor and sweetness, but they have a different effect than honey because of the crystalline nature of sucrose, which is the main component in brown sugar. The sweeteners in various confectionery products are listed in Table 7-1.

Confections can be divided into several general categories depending on the manufacturing procedures and ingredients used. The products in these groupings tend to have similar characteristics and features, although several formulation variations may exist for a given confection. For a more detailed explanation of confection types than is given in this chapter, the reader should consult a specialized reference.

Functions of Sweeteners in Confections

SWEETNESS

The primary function of sweeteners is to provide sweetness. The choice of a sweetener or mix of sweeteners determines how sweet the product will be. Sweetness as a sensory property is discussed in Chapter 2.

TEXTURE

Textures of confectionery products include hardness, softness, and chewiness, as well as body or mouthfeel.

The single most important ingredient influencing the texture of confections is sucrose because much of candymaking involves the dissolving and subsequent recrystallization of sucrose (1). Supersaturated solutions are formed and then cooled to cause recrystallization or solidification (2).

Crystallization of a sugar solution is known as *graining*. If the sugar solution is allowed to cool in a slow, undisturbed manner, large crystals are obtained, as in rock candy. If cooling is rapid and with agitation, the result is fine crystals, as in cream centers or fondants.

If and how sucrose solidifies is also controlled by the second largest ingredient in confections, corn syrup. Higher molecular weight sugars, particularly in lower-DE products, inhibit the crystallization of sucrose and other simple sugars. Thus, hard candy, which contains glucose syrup, is a solid but is not crystalline. Corn syrup also provides a soft texture to caramels, marshmallows, and gummed candies and helps stabilize foams in marshmallows and nougats.

Graining—Forming a crystalline structure.

TABLE 7-1. Typical Sweetener Composition (%) of Confections

Product	Sweetener		
	Sucrose	Corn Syrup	Other
Milk chocolate	35	...	30 NFDM or milk solids
Hard candy[a]	47	37	
Lollipops	67	17	
Peanut brittle	42	21	
White taffy	61	15	
Fudge			
Chocolate	60	1	
Vanilla	35	24[b]	12 fondant sugar
Creole pralines	52.5	...	
Caramel	15.4	34.5[c]	3.8 dextrose
Caramel/fudge[d]	30–60	20–50	0–6 NFDM or milk solids, 1–10 invert sugar
Gummed candies[d]			
Pectin jellies	45–55	20–30	
Agar jellies	45–55	20–30	
Starch jellies	30–45	25–40	
Gum arabic			
Soft	15–35	15–35	
Hard	20–35	10–20	
Starch gums			
Soft	25–40	20–35	
Hard	20–45	15–35	
Marshmallows[d]	55–60	20–25	
Nougats	33	21[e]	11 honey
Licorice[d]	10–20	20–50	
Citrus tablets	...	...	90 dextrose
Butterscotch topping	...	82.0[f]	2.4 brown sugar

[a] May contain other minor sweeteners such as molasses, lactose, or fruit juice (fructose).
[b] 24-DE corn syrup preferred.
[c] 42-DE corn syrup preferred.
[d] Data from (1).
[e] 62-DE corn syrup preferred.
[f] 42–62 DE.

HUMECTANCY

Some of the ingredients must be hygroscopic, attracting and retaining the water molecules present in the ingredients so that the final products don't dry out. Sweeteners like sucrose, invert sugar, fructose, and dextrose can act as humectants. The higher molecular weight sugars in products like corn syrups are less hygroscopic, and they help prevent stickiness caused by too much water absorption.

Processing Considerations

Candymaking is often thought of as an art. There are many skilled confectioners who know just when to stop pulling a taffy so that it doesn't become too hard or when to add more frappé to a nougat in order to get the right final stiffness. However, understanding the science behind the artistic manipulations can help to clarify candy manufacturing. Candymaking can also be defined as a manipulation of the physical state of the sugars present, namely, the crystalline form of the sweetener molecule. In general terms, sugars melt upon heating, and, when subsequently cooled, they recrystallize. The crystals that form during cooling depend on many factors, including agitation; temperature (i.e., boiling and cooling); viscosity; type of sweetener used; presence of other ingredients such as gums, starches, fats, etc.; and the pH of the system.

AGITATION

The more a molten sugar solution is agitated or beaten, the greater the number of crystals that form during cooling, resulting in smaller and more numerous crystals. This is important in the manufacture of a smooth *fondant* or creme. Likewise, less agitation or beating results in fewer, larger crystals and a coarser, grainier confection. This is considered a defect in, for example, a nougat but is necessary for making rock candy.

TEMPERATURE

Another factor that affects the final state of the crystal form is the temperature to which the formulation is heated as well as the time at which it is held there. In general, the higher the temperature and, if boiled, the longer the boiling time, the more the initial crystals dissolve and melt and thereby have the ability to recrystallize. If not all crystals are allowed to dissolve and melt, they act as nuclei or seeding sites on which new crystals form and grow, and therefore, fewer, larger crystals form. In the making of a hard candy, if not all of the sucrose is dissolved in the solution (e.g., if some remains on the rim of the pot or kettle and is then wiped in with a spatula), a grainy, cloudy, poor-quality candy results.

Cooling is also important. In hard candy manufacture, the cooling conditions and formulation are such that the molten solution of sugars does not recrystallize but forms an amorphous glass, which can be seen in many of the candies in Figure 7-1. In the manufacture of caramels, where graininess due to sugar recrystallization is considered a defect, the mixture must cool and set without recrystallizing.

VISCOSITY

Viscosity has a large effect on the crystal forms of a final confection. The viscosity depends mainly on the formulation's mois-

Fondant—A grained confection often used as an ingredient in the manufacture of other candies such as fudge.

ture content. However, other ingredients such as fat sources can influence the product's viscosity. Also, higher-DE corn syrups increase the fluidity of a viscous formulation, while lower-DE syrups increase the viscosity. The more viscous a formulation, the more difficult it is for sugars to recrystallize and grow. Many different textures of candy can be obtained by varying the viscosity of the formulation.

SWEETENER TYPE

Sucrose has the ability to crystallize easily, while fructose crystallization is more difficult. The differences in their chemical structure affect their ability to attract water, i.e., to hydrate and become soluble. At a given temperature and concentration, fructose is generally more soluble than sucrose. Fructose, therefore, tends to stay in the syrup phase of a formulation, while sucrose, depending on the conditions (i.e., moisture content), is more likely to crystallize into a solid form. Honey, because it is comprised mainly of fructose and glucose, tends to remain in the syrup phase rather than crystallize. Corn syrups and honey are often used to retard sucrose crystallization in candies such as caramels and chewy toffees.

Fig. 7-1. An assortment of candies.

OTHER INGREDIENTS

The presence of other ingredients also affects a sweetener's ability to form a crystal and grow. Ingredients such as starches, gums, or proteins interrupt the crystal nucleus sites where the sugar molecules can migrate to grow. Therefore, depending on the formulation and processing conditions, the sweetener may remain in a dissolved syrup state, or smaller crystals may form but not grow. In caramel manufacture, the presence of milk proteins helps to create smooth final products by inhibiting crystallization.

Acids also affect crystallization. If the pH is sufficiently low, sucrose can hydrolyze or invert to its component sugars fructose and glucose, which retard crystallization. Cream of tartar is an acidulant often used to achieve this effect.

Chocolate and Compound Coatings

The basic ingredients in chocolate and coating formulas include cocoa nibs, refined sugar, cocoa butter, milk, butter oil, and flavor. Cocoa itself is very bitter to the taste. The main sweetener in chocolate products, *compound coatings*, and cocoa drinks is sucrose, which can comprise up to 50% of the formula on a weight basis. It is used because it has low hygroscopicity and is easy to blend into the formula and process. Lactose, which is also often present in products such as milk chocolate, is not usually added as an ingredient for its sweetening or other functional characteristics but is present naturally in the dairy-based ingredients used. Some formulations substitute corn syrups for sugar as a cost savings as well as to decrease the overall sweetness.

Many varieties of chocolates and coatings are available in the market. There are dark chocolate and milk chocolate processes and also formulations for white chocolate (essentially chocolate without the cocoa) and other compound coatings.

PROCESSING

The basic procedure for chocolate manufacture can be outlined in the following steps: preliminary mixing, refining, *conching*, tempering, and final product preparation, which may include panning, *enrobing*, and/or molding. The final step is the cooling of the product. Figure 7-2 shows the steps for processing dark and milk chocolate.

Sucrose used in chocolate manufacture must be dry to eliminate caking and must have a consistent particle size. In addition, the sucrose must not contain invert sugar (i.e., sources of glucose and fructose, which would decrease viscosity).

Sugar problems usually appear first in the refining step. Refining machines are essentially rollers through which the premixed ingredients pass. The end product of refining is a smooth, pastelike chocolate material. If the refining step is not done properly or if the sugar is not dry or the particle size is not consistent, the paste does not have a smooth consistency, and a defective product results.

The conching step is responsible for the final mixing as well as for the flavor development and final texture of the chocolate product. The subsequent steps of molding and cooling result in the final product shape and appearance. If invert sugar is present, the viscosity of the chocolate may be too thin and it may not set properly in the mold or coat properly during enrobing.

BLOOM

Bloom, a well-known defect in chocolate products, appears as white or gray flecks on the surface of the chocolate and is not a health issue. The two types of bloom are fat bloom and *sugar bloom*. Fat bloom occurs when the incorrect crystalline form of the fat is present in the final product. Sugar bloom is caused by moisture migration rather

Compound coatings— Coatings containing fats other than cocoa butter but similar to regular chocolate in melting properties.

Conching—Slow mixing of a heated chocolate paste to reduce particle size and increase thickness and smoothness.

Enrobing—Covering a base food material with a melted coating that hardens to form a solid surrounding layer.

Sugar bloom—A dusty white appearance on the surface of chocolate caused by the formation of certain types of sugar crystals.

than by crystallization changes. It typically has a grayish appearance and cannot be easily removed from the surface of the chocolate. It also does not have a greasy feeling, as fat bloom does. The two types of bloom are most easily distinguished by looking at the defect with a microscope. If the appearance of the bloom is rough and crystalline, it is sugar bloom.

Sugar bloom can result from a number of moisture problems. In a humid environment, the sugar in the chocolate absorbs water. When the candy is moved into a drier environment, the moisture on the surface evaporates and the sucrose recrystallizes on the surface. In the early stages of bloom, a sticky syrup often appears on the candy surface. Sugar bloom is therefore caused by damp storage; damp cooler air; impure, hygroscopic, or brown sugars; or improper packaging. It also results when chocolate is coated onto a particulate or center that has a higher equilibrium relative humidity (or higher water activity) than the chocolate; the moisture is trapped by the wrapping and causes bloom when the product is held at a higher temperature. Once sugar bloom occurs, there are no real solutions, as the sugar that crystallizes onto the surface was once an integral part of the chocolate product.

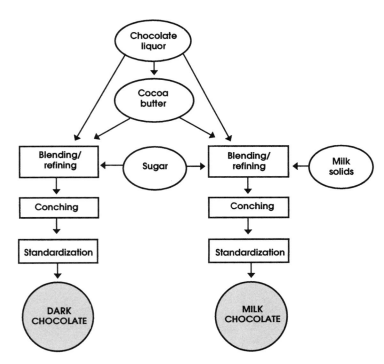

Fig. 7-2. Steps in the manufacture of dark and milk chocolate. Adapted from (3).

Hard Candies

Hard candies are based on the principle that sugars can form an amorphous glass. Hard candy formulations are relatively simple, comprising sugar, a glucose syrup, color, and flavor. By itself, sugar dissolved in water cannot be used to make hard candy because it crystallizes too easily, resulting in a defective product. In the past, confectioners used cream of tartar in their formulations to produce invert sugar, which retarded crystallization. However, this practice produced inconsistent product from batch to batch since the amount of invert sugar produced depended also on the hardness of the water and the purity of the sucrose. Today, manufacturers use glucose-containing syrups, which produce a hard candy that is less *short* in texture and less sweet than that made with invert sugar. Glucose-containing syrups, usually used at a 20–50% range in the formula, add body to the end product. Generally, syrups with a DE of 38–48

Short—Describing the texture of a product that breaks apart very easily when bitten.

are used in hard candies. They help enhance flavors, maintain good moisture control, and assist in emulsifying colors and flavors.

PROCESSING

The making of hard candy begins with the mixing of the sucrose, syrup, and water. It is important that the sugar be free from any caking or lumping and be of uniform particle size. The mixture is brought to boiling, which usually is in the range of 151.7–168.4°C (305–335°F). If the crystals do not completely dissolve or if any remain on the sides of the kettle, they act as sites for nucleation and crystal growth. A high-viscosity syrup is desired at boiling. If the viscosity is too low, crystallization occurs. The viscosity of the sucrose-water solution is increased by the addition of a low-DE glucose-containing syrup.

The three common methods of boiling hard candies are 1) open pan, 2) vacuum, and 3) continuous film or a scraped-surface heat exchanger. The ratio of sucrose to glucose in open-pan boiling is about 70:30, while in vacuum boiling it can be 60:40. Continuous-film boiling uses a ratio of about 65:35. Vacuum boilers are often used in clear or colorless candies since they decrease the temperature needed to boil and therefore produce less browning or off colors in the final product. High-maltose syrups are employed in processes requiring extremely high temperatures. They have less reducing sugar and produce fewer problems with browning. They also tend to form more brittle candies, so processing temperatures (and costs) can be reduced.

Colors are usually added during boiling. The solution is then cooled and poured onto an oiled table, where flavors are added and kneaded into the molten mass. The still warm and plastic mass is then either molded, rolled, or machined into its final form. When cooling, the sugars do not crystallize but form a glass.

The mass can also be pulled to form an opaque and translucent medium due to the incorporation of air. This results in the crystallization of sucrose. After shaping and cooling, the confection has a spongy, shorter texture due to the crystallization that has occurred.

DEFECTS

Some key problems that occur in hard candies revolve around sugars. The sugars can invert due to addition of acids such as citric, malic, or tartaric acid, which are used for flavor. This causes the candy to become hygroscopic and sticky. Vacuum boiling reduces the inversion because of its ability to boil solutions at decreased temperatures and for less time. Sticky candies can also result from improper packaging material or from having exposed candy (e.g., bulk or binned candy) in a humid room.

Another defect, already mentioned, is graining or sucrose crystallization in the candy. This can happen in a variety of ways. Crystallization can occur if the boiling process is incomplete or if sucrose remains on the kettle rim. It can also occur when the sucrose-glucose ratio is too high. In general, the ratio should not be higher than 75:25. If it is too high, increasing the glucose syrup helps to eliminate the problem. Crystallization can also occur if the final moisture of the candy is higher than about 1–2%. Humid storage conditions are another cause. Sucrose solubilizes in the humid air on the surface of the candy and then recrystallizes on the surface when the surrounding atmosphere becomes drier. The addition of reworked candy to the formulation or process can also cause crystallization, so care should be taken to adjust the processing temperatures and/or times to allow complete dissolving of sucrose.

Caramels and Other Chewy Candies

Chewy candies are essentially the next formulation step from hard candies in that the glucose or corn syrups are increased in proportion to sucrose. Chewy candies are cooked at lower temperatures than hard candies and have a higher final moisture content. They do not contain crystallized sugars. The quality of chewy toffees, taffies, and caramels often depends on their smoothness.

Caramels are a special subset of the chewy candy group in that they incorporate proteins from dairy-based ingredients such as sweetened condensed milk, whey powders, cream, or skim milk. These proteins react with the reducing sugars present, forming the pigments and distinctive flavor associated with caramels. Because milk proteins have a tendency to scorch, lower temperatures are used (about 115.6–121.1°C or 240–250°F) for cooking and the milk ingredients are usually added late in the cooking process.

Toffee is similar to caramels except that it is boiled at higher temperatures (about 148.9°C or 300°F) and has a lower amount of dairy ingredients and less fat. The resulting final product is therefore harder (due to the lower moisture content and fat) and, in general, is darker in color.

As with hard candy, chewy candies can be pulled to incorporate air. The most familiar of these pulled chewy candies is taffy.

Several ingredients not found in hard candies may be added to chewy candies. Lower-DE syrups are used to increase the viscosity of the formula and, in general, produce a tougher, less sweet product. Likewise, the use of higher-DE syrups decreases the viscosity of the formula and results in a softer, sweeter, and possibly darker product, depending on cooking time and temperatures. Fats from butter and cream also help to soften caramels and other chewy candies and act as release agents that keep the candy from sticking to the teeth.

Gummed candies include such familiar products as jelly beans, gumdrops, fruit slices, and gummy bears. They can be made with gums (pectins or agar) or with starch, which are used to help set the shape and texture of products. The softer candies are generally high in sucrose, medium in corn syrup, and low in starch or gum. The stiffer ones have higher viscosity because they are high in starch or gum and medium in sucrose and corn syrup. Regular corn syrups are usually used because they control crystallization and hygroscopicity. Some gummed candies are made with 40% honey.

Fudge

Fudge is essentially a chewy candy that has been grained, or crystallized. In fact, the story of the making of fudge is something of a legend. There are several versions of the story, but all agree that fudge was the product of a frugal caramel maker. The confectioner's caramel batch went awry when it was overcooked while being continuously stirred. Because of the overcooking and overstirring, the caramel batch had darkened and grained. After tasting the caramel, the confectioner thought that the "ruined" concoction tasted quite good. Today, fudge is manufactured more consistently, using a caramel base to which a fondant is added. Fondant, described in more detail in the next section, can be simply defined as a grained confection used to cause the crystallization of the fudge's caramel base.

The are many varieties of fudge ranging from firm, short textures to softer, more moist textures. The final flavor and texture of fudge depends on how much the caramel base is boiled, the amount of added fondant, and the proportion of sucrose to glucose in the caramel base formula. In general, higher boiling temperatures, longer boiling times, more fondant, and higher sucrose ratios lead to faster crystallization of sugars in the fudge and firmer, shorter textures. The temperature at which the fondant is added is also important. If it is too high (in general, higher than 51.7°C [125°F]), the texture will be coarse and gritty. If it is too low (less than 32.2°C [90°F]), the fudge may not grain at all.

Fudge can be produced either by batch or continuous processes. Fudges produced via batch process tend to have a shorter texture than those made by continuous process. When using a continuous method, it is better to use higher ratios of sucrose to glucose (about 10:1) to aid crystallization rather than the typical 4:1 or 3:1 ratio of a caramel formula.

Fondants and Cremes

Fondants are used as both an ingredient (e.g., in fudge) and as a final product. The term "creme" is often used interchangeably with "fondant" but usually denotes a fondant that is made up of sugars only. Fondants can be varied and can contain other miscellaneous ingredients such as flavors, colors, or nuts.

Fondants are made by mixing a supersaturated sugar solution (about 12% water) that is usually sucrose with some invert sugar or glucose-containing syrup. A combination of lactose and honey can be used instead. The solution is brought to a boil via open fire kettle, steam-jacketed kettle, or vacuum boiler. Then, in the creaming step,

Typical Temperatures for Candymaking		
°C	°F	Stage
110	230	Syrup
116	240	Fudge (soft ball)
121	250	Firm ball
127	260	Hard ball
138	280	Soft crack
149	300	Hard crack

it is cooled with intense mixing or beating, during which minute sugar crystals form, suspended in a liquid syrup phase.

As with hard candies, earlier formulations included cream of tartar to create invert sugar in the mix to make the syrup phase. This produced very sweet fondants with short textures. Inconsistent products were also common because the hardness of the water affected the boiling time and temperatures. Today, glucose syrups are used to produce consistent fondants that are less sweet and have a texture that is not as short. If too much glucose is added, however, needle-like crystals can form, causing an undesirable waxy texture. The sucrose-glucose ratio is typically about 80:20. Fondants with a low sucrose-glucose ratio are not only too fluid but are also susceptible to microbial problems. The moisture content of fondants is roughly 12%. If the moisture content reaches 15–20%, microbial growth may occur.

The quality of the fondant depends on several factors. Speed and efficiency of beating are very important in generating numerous small crystals. The temperature at beating initiation, i.e., the creaming temperature, is also a key factor. If the temperature is sufficiently high, larger sugar crystals result, but if it is too high, the texture could be gritty. If the temperature is sufficiently low, smaller crystals are produced. The optimum crystal size for fondants is between 15 and 20 μm. Retardation of crystallization and crystal growth is important for the quality of the fondant over its shelf life. Glucose syrups, starch, gelatin, and proteins help to maintain the quality of the fondant by retarding crystal formation and growth.

Fondants, once prepared, are often cooled, ripened or matured, and remelted. There is debate among confectioners about whether the ripening/maturing step is necessary or provides any additional benefits. Remelting is used when making creme centers. Cooled fondants are heated to 57.3–82.2°C (135–180°F), and a syrup "bob" is added to the reheated mixture. The bob syrup is a formulation that is similar to the original fondant formulation but contains a higher proportion of a glucose-containing syrup. The crystal size after the remelt stage and subsequent cooling stage depends on the temperature and speed of those two stages. In general, temperatures over 65°C (153°F) tend to produce coarser textures because of the larger crystal sizes that result. Confectioners can therefore tailor-make the final fondant by manipulating these parameters.

Aerated Candies

Candies can be aerated by the addition of air through pulling (mentioned above for hard and chewy candies), by chemical means (such as the addition of sodium bicarbonate), or by the addition of a frappé. A frappé is essentially a fondant to which a whipping agent has been added. Whipping agents are usually egg albumen, gelatin, or a hydrolyzed soy protein. The whipping agent is dissolved in water and then added to the fondant while it is being whipped to a foam. The final frappé depends on the level of whipping agent added and the time and speed of whipping. In general, the more syrup present in the initial fondant, the denser the final frappé. If the syrup phase is too high, however, a coarse, gritty texture results.

A nougat is a fondant that has been further aerated with a frappé. Fat is also added to nougat formulations to decrease the stickiness and act as a release aid when cutting the final product. The quality of the frappé as well as the fondant is important to the final quality of the nougat. Shortness, degree of aeration, and density of the nougat can all be tailored by adjusting the frappé characteristics and the fondant quality.

A marshmallow can be defined as a nougat without the fat. It is also higher in moisture than a nougat. Marshmallows are prepared by adding an aerating agent to sucrose and glucose syrups. The mixture can then be poured into starch molds or extruded into various shapes. The final texture of a marshmallow depends on its moisture content (which can range from about 12 to 18%), the amount of whipping agent added, and the final processing or shaping step. The two main defects that can occur include graining and dehydration. Ingredients such as glycerol or sorbitol, agar, gum arabic, or pectins are often added to help minimize these defects and maintain product softness.

Others

CITRUS-FLAVORED TABLETS

The manufacture of candy tablets is very similar to that of pharmaceutical tablets. The main difference is that candymakers prefer to use the sweeter varieties of tableting materials. Among the sweeteners, there are tableting grades of sucrose, dextrose, and lactose. However, in most confectionery applications, dextrose is the preferred material for several reasons. First, dextrose gives the hardest tablets. If needed, they can be softened with small amounts of ordinary grades of dextrose, sucrose, or some other sweetener. Second, dextrose is more stable under the acidic conditions of the citrus-based products, since food acids do not cause inversion as they do with sucrose. Third, the negative heat of solution with dextrose causes a cooling effect, which is complementary to citrus-flavored candies. Other sweeteners, such as maltodextrins used as flavor carriers, may also be present in small amounts.

PANCOATED CANDIES

Panning is a process in which a thin layer or shell is applied to a candy center while it is tumbling in a large rotating pan or vessel. Its most familiar use is for candy-coated chocolate pieces. The main component of a pancoating is sucrose (4), which by itself forms hard panned items. Ingredients such as corn syrup and chocolate are added to make softer and chocolate-flavored products, respectively. Dextrose has also been used as the principal component and reportedly reduces panning and drying time.

The shelf life of pancoated confections is improved by sealing the centers with starches, gum arabic, or other carbohydrates. These coatings make the surface of the candy center smooth and receptive to panning layers, and they also act as a moisture and fat barrier. Maltodextrins are also used in some pancoatings because they have low sweetness and good binding and film-forming properties. Corn syrup solids are also used for the same reasons and to add sheen to certain panned candies. Good film-forming products provide good protective coatings.

Troubleshooting

CHOCOLATE AND COMPOUND COATINGS		
Symptoms	**Causes**	**Changes to Make**
White color on surface	Sugar recrystallization on surface	Monitor humidity during cooling and storage. Use proper packaging.
	Surface moisture on candy to be enrobed	Thaw frozen pieces completely, in a low-humidity at atmosphere.
		Check that temperature of cooled piece is above dew point of ambient air temperature and relative humidity.
Consistency too thin or doesn't set properly	Moisture content too high	Check formulation and/or processing conditions. Ensure adequate mixing. Decrease levels of invert sugars. Increase pH during processing.
Grainy or gritty texture or crumbly structure	Moisture content too low or improper sugar crystal formation	Check processing conditions to ensure proper crystal formation. Check particle size and uniformity of sweeteners.
FUDGE		
Symptoms	**Causes**	**Changes to Make**
Consistency too thin	Insufficient crystallization or moisture content too high	Increase cooking time and/or temperature. Increase agitation. Increase sucrose-glucose ratio. Check temperature at which fondant is added to caramel base.
Coarse or gritty texture or crumbly structure	Low moisture content or large sugar crystal formation	Decrease cooking temperature and/or time. Decrease sucrose-glucose ratio. Check temperatures during processing. Ensure proper cooling before agitation.

CARAMEL AND CHEWY CANDIES

Symptoms	Causes	Changes to Make
Color too dark and/or scorched flavor	Excessive Maillard browning or caramelization	Decrease reducing sugars. Use high-maltose syrup with high processing temperatures. Use corn syrups with lower DE. Decrease milk solids. Decrease boiling temperature and/or time.
Lack of color and/or flavor development	Insufficient Maillard browning or caramelization	Increase reducing sugars. Increase milk solids content. Use corn syrups with higher DE. Increase boiling temperature and/or time.
Gritty or grainy texture	Sugar crystal formation	Decrease cooking temperature and/or time. Decrease sucrose-glucose ratio. Prevent agitation during cooling.

HARD CANDY

Symptoms	Causes	Changes to Make
Gritty, grainy, or short texture	Sugar crystal formation	Reduce sucrose-glucose ratio to less than 75:25. Limit amount of added rework. Increase boiling time or temperature to lower moisture content (to 1–2%). Check quality of incoming ingredients. Add acid if using sucrose as sole sweetener.
Product too sticky	Moisture content too high or crystals not in amorphous state	Increase sucrose-glucose ratio or add maltodextrins. Monitor pH to control sucrose inversion. Increase boiling time or temperature. Cool rapidly without agitation.

GUMMED CANDIES

Symptoms	Causes	Changes to Make
Product too soft or too sticky	Moisture content too high	Increase sucrose-glucose ratio. Check processing conditions.
Product too stiff or hard	Moisture content too low or water too tightly bound	Increase ratio of glucose-containing syrup to thickener. Decrease thickener content.

MARSHMALLOWS AND NOUGATS

Symptoms	Causes	Changes to Make
Product hardens or toughens	Excessive drying	Increase glucose-sucrose ratio. Replace some sucrose with invert sugar or HFCS. Add humectant agents.
Coarse and/or gritty texture	Sugar recrystallization	Decrease sucrose-glucose ratio.
Product too dense	Foam breakdown or lack of foam development	Decrease syrup content.

FONDANTS AND CREMES

Symptoms	Causes	Changes to Make
Waxy texture	Formation of needle-like crystals	Decrease glucose content.
Product too runny or sticky	Insufficient crystallization	Increase agitation. Increase cooking temperature and/or time. Increase sucrose-glucose ratio.
Gritty or grainy texture	Formation of large crystals	Increase agitation. Decrease creaming temperature.
	Crystal growth during storage	Add starch, gelatin, or protein to retard crystallization.

References

1. M. Barnett. 1990. Sugar in confectionery. Pages 103-129 in: *Sugar: User's Guide to Sucrose*. N. L. Pennington and C. W. Baker, eds. Van Nostrand Reinhold, New York.
2. Hegenbart, S. 1995. *Food Product Design*, April, pp. 28-44.
3. Chandran, R. 1997. *Dairy-Based Ingredients*. American Association of Cereal Chemists, St. Paul, MN.
4. Rapaille, A., Gonze, M., and Van der Schueren, F. 1995. Food Technology 49(7):51-54.

Other Applications

Fruit and Vegetable Preservation

Fruits and vegetables are preserved mainly by drying, freezing, and canning. Syrups and granular dry sweeteners are used in these processes. Although sweeteners are more common in the preservation of fruits (i.e., in jams and jellies), many preserved vegetables also contain sweeteners that help to create desired product attributes.

EFFECTS OF CARBOHYDRATE-BASED SWEETENERS

Flavor. Carbohydrate-based sweeteners enhance the flavors of foods by several means. The addition of a sweetener enhances sweet flavors and also reduces or tames acid notes, as in the addition of sucrose to canned tomatoes or tomato pastes. Sour or bitter notes can also be decreased with the addition of sucrose (e.g., in the canning of cherries). Some types of cans give a metallic flavor that can be avoided by the use of ion-exchanged corn syrups.

Texture. Simple sugars such as glucose, fructose, and sucrose are able to diffuse across cell membranes of plant tissues, thus increasing osmotic pressure, as discussed in Chapter 5. In general, the osmotic pressure increases with decreasing dextrose equivalent (DE). The increase in osmotic pressure helps to keep the tissue firm, which is important for freezing fruits and vegetables. Addition of low molecular weight sugars decreases the freezing point of the tissue, which causes less ice formation within the tissue and thus less tissue damage. High molecular weight components, present in corn syrup and maltodextrins, control the presence of ice crystals and the size of sugar crystals.

The syrups of canned fruits and vegetables also benefit from the addition of sweeteners because the sweetener increases mouthfeel and body as well as adding a gloss or sheen to the product. Corn syrups are often used to create a thick syrup.

Preservation. Sweeteners preserve fruits and vegetables by lowering the water activity of the food, making water less available to spoilage microorganisms, yeasts, and molds. This is important in the drying or dehydration of foods. Some water remains in the tissue so that it does not become overly hard, yet the water is preferentially bound to the sweetener. Sweeteners in syrup form also protect frozen and canned foods from oxidation. The syrup forms a protective coating around the food, thereby reducing the diffusion of oxygen into the tissues.

JAMS, JELLIES, and PRESERVES

Jams, jellies, and preserves are basically made by the same procedure, with the form of the main fruit ingredient being the distinguishing difference. Jellies are made with the juice extract of the fruit and do not contain actual fruit pieces. Jams start with the crushed fruit, and pieces of the fruit are part of the final product. Preserves are made with crushed or entirely whole fruit. The U.S. Code of Federal Regulations contains *Standards of Identity* that list the requirements for the fruit of these products. For this discussion, the term "jam" is used to refer to all characteristics of jams, jellies, and preserves unless otherwise noted.

In general, the process for making jam is to combine fruit and sugar and bring them to a boil either in a vacuum kettle or an open pan to dissolve the sugar. Although fruits naturally contain pectin, not enough is present to form a gel structure, especially in the making of jellies, where the pectin-containing components are removed. Commercially available pectins are therefore added to ensure gel formation. The mixture is then packaged, cooled, and allowed to set.

The final product is about 45% by weight fruit or juice and about 55% sugar. The amount of soluble solids (measured by a refractometer) is roughly 65–70%. The term "soluble solids" is used because sugars other than sucrose (such as fructose and glucose, which are naturally present in fruits) are also measured. If the jam is used as a filling in a confection, the soluble solids are generally higher (about 75–78%) to help reduce the risk of microbial spoilage. If the soluble solids are higher than 78%, the sucrose recrystallizes, resulting in a defective product.

The sweetener used to make jam is generally sucrose, which imparts a clean, sweet taste and a firm gel to the final product. Honey and high-fructose corn syrup (HFCS) can also be added to aid the retardation of sucrose crystallization in the final product. Corn syrups are used less often because they lower the product's sweetness level. About 50% of the sucrose can be replaced, but a decrease in the gel strength can occur if the sweetener is less than 50% sucrose. Also, since syrups contain a higher proportion of water, the formulation must often be boiled longer to remove the excess water. Upon storage, however, the fructose and glucose in the product help to retain the moisture, often leading to a further loss of volatiles and a lower-quality product. High-quality jams most often use sucrose as their sole added sweetener.

A wide array of pectins are available for jam making and are chosen by the type of jam qualities desired. Pectin is a polymer of D-galacturonic acid. The acid groups can also be methoxylated. Figure 8-1 shows a portion of a pectin molecule that contains both acidic groups and methoxylated groups. Pectin gelation occurs when the pectin molecules align to form junction zones, which are matrices that include the dissolved sugars and fruit components. The gelation depends on the state of the groups (methoxylated or in their acidic form) and on the pH and amount of sugar present. The sugar attracts the water, thus allowing hydrogen bonding between the pectin molecules, which forms the junction zones. If the pH is low and the groups are in their nonmethoxylated (acidic) form, the hydrogen bonding is also enhanced and pectin gelation can occur. If the pH is too high, the acidic groups repel each other and do not form junction zones.

Gelation of high-methoxy pectins requires sugar and acidic conditions in the system. The pH range needed for gelation of high-methoxy pectins is about 2.9–3.6, and the soluble solids range is about 60–80%. If the pH is too low, *syneresis* of the gel can occur and syrup will weep out of the gel over time. The pectins also must be fully dispersed in the boiled fruit medium and not be degraded by overcooking. If gelling occurs before the product is poured into its final container, a rough, grainy product can result.

Low-methoxy pectins, on the other hand, do not need acidic conditions or sugar for gelation. They depend on the presence of a calcium salt such as calcium citrate. Calcium forms a bridge between the two pectin molecules, thus forming a junction zone. These types of pectins form rapid and somewhat irregular gel structures and are often called "rapid-set" pectins.

Sucrose inversion occurs during storage because of the acidic environment of most final products. Sucrose inversion can also occur if the formulation is boiled for too long or is insufficiently cooled. If sucrose inverts to such an extent that the soluble solids increase, the sucrose may recrystallize, making a defective product. Also, if the dextrose concentration is greater than 35%, it can crystallize. To keep this problem from occurring, high-maltose syrups are sometimes used.

CANNED FRUITS AND VEGETABLES

Some canned vegetables, like beans and mixed vegetables, use little or no sweeteners. On the other hand, corn, creamed corn, beets, and peas contain some sugar. Canned fruits such as pears, peaches, fruit cocktail, and apple sauces use corn syrup and either sucrose or HFCS, but mandarin oranges and some fruit cocktails contain just sucrose. Most canned pie and pastry fillings use corn syrup, while

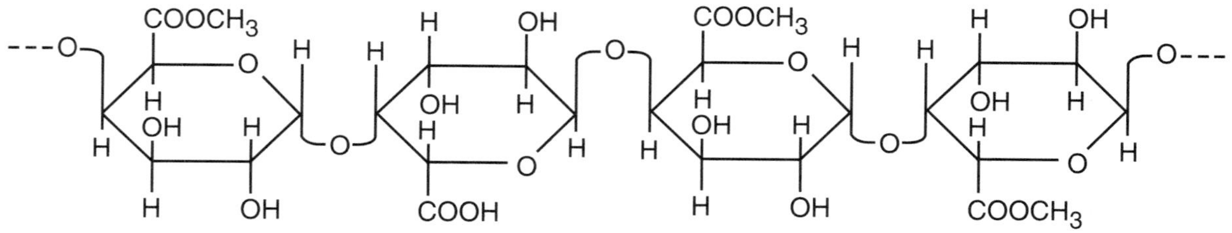

Syneresis—The separation of a liquid from a gel; weeping.

Fig. 8-1. Pectin molecule containing both acidic groups and methoxylated groups.

lemon pie fillings generally contain sucrose, and apple fillings use HFCS.

In canning applications, HFCS is frequently used in combination with corn syrup. Because of its high osmotic pressure, it rapidly penetrates fruit and vegetable tissues with a minimum of tissue damage. In addition to being a sweetener, HFCS acts as a preservative and adds sheen. Fructose is especially good in canned fruits and fruit fillings because it enhances flavor. High-DE (54–62) syrups are used in canned fruits for sweetness, sheen, and the ability to lower water activity, which helps reduce the growth of microorganisms.

Depending on the type of can employed, some canners have found that ion-exchanged corn syrups are needed to avoid metallic flavors in canned fruits. The chloride ion concentration must be very low when certain two-metal cans are used.

Beverages

Typical sweetened beverages available in the marketplace are of several types: carbonated; fruit, vegetable, or dairy-based; powdered; and alcoholic. Most beverages are sweetened with either sucrose or HFCS (usually 88% fructose). Naturally present carbohydrate-based sweeteners such as maltose and lactose are also common in alcoholic and dairy-based beverages, respectively. The formulation and processing steps depend on the beverage type.

CARBONATED AND NONCARBONATED BEVERAGES

The most familiar carbonated beverage has several familiar names such as soft drink, pop, or soda. In general, carbonated beverage formulations are about 90% water and 10% sweetener. Minor amounts of color, carbon dioxide, flavor, acidulant, or preservative are also often added.

Sucrose was once the most widely used ingredient in this industry and is still used in many formulations. Corn syrups and invert syrups were tried for a while as sucrose replacements to reduce the cost of beverage formulations. These sweeteners were not as effective because they have limited sweetening power (i.e., low sweetness compared to sucrose). In the late 1970s and early 1980s, HFCSs began to be utilized because of their availability, low cost, and ability to maintain sweetness when replacing sucrose. Today, numerous carbonated drinks contain HFCS. Although the 42%-fructose HFCS has the same sweetness as sucrose, the sweeter 55% product must be used (Table 8-1). This is because, in sucrose-sweetened carbonated beverages, part of the sucrose is converted to invert sugar in the low-pH environment and the mixture is sweeter than would be predicted from the sucrose alone.

Sucrose concentration is used during the production of carbonated beverages as a quality control measure to ensure a consistent product. The percent sucrose (on a weight basis) present in a sugar-water solution (often reported as degrees Brix) can be measured using a hydrometer or a refractometer (see Chapter 4). In a formulation with sucrose as the only sweetener, the Brix reading is directly proportional to the amount of sucrose in solution. However, over time, sucrose can invert to its component sugars, glucose and fructose, which affects the final taste and sweetness of the formulation. In addition, the true percent solids increases, as does the volume of the beverage. The Brix reading, therefore, is lower than the true solids value since Brix measures only sucrose. Brix correction charts have been developed to account for the presence of invert sugars in solutions. They are useful when replacing sucrose with invert syrups or other glucose- and fructose-containing syrups such as HFCS.

Several factors affect sucrose inversion. The lower the pH of the beverage and the higher the processing or storage temperature, the faster the inversion. However, inversion of sucrose may be a deliberate processing step. In this case, inversion may be induced by an acidulant such as citric acid with or without added heat. The enzyme invertase may also be used.

TABLE 8-1. Concentration (%) of Sweeteners in Liquid Beverages

Product	Sucrose	Fructose	HFCS,[a] 55%	Aspartame
Cola				
Regular			15.1	
Lite		3.5		0.025
Zero-cal				0.077
Fruit drink				
Regular	8.3			
Lite		5.7		
Lemonade				
Regular	22.9			
Lite		17.5		

[a] High-fructose corn syrup.

Noncarbonated beverages include those made from fruits, vegetables, and dairy-based ingredients. The formulations vary widely. Sweeteners may not be necessary because of the natural presence of sweeteners such as fructose, glucose, and lactose from the primary ingredient. However, variation in the crop or source can cause inconsistencies in final-product sweetness, so sweeteners are often added to ensure a consistent final product (Table 8-1). Sucrose or HFCS, as described above for carbonated beverage production, are the sweeteners most often used. Corn syrups or maltodextrins are also used in noncarbonated beverages to build body in the product and help provide stability and emulsification. They are used when low levels of added sweetness are desired.

POWDERED DRINK MIXES

Carbohydrate-based sweeteners in powdered drink mixes must add sweetness to the final product without settling to the bottom of the container. To avoid segregation problems, fine granulations of sugars are used. Sucrose is the most common sweetener in these types of products, although dextrose, corn syrup solids, or maltodextrins can be used to increase the mouthfeel and body of the reconstituted product. The choice of bodying agent depends on the final sweetness level desired. Often several carbohydrate-based sweeteners are used to create a sweet-flavored product with body and texture. Dextrose can replace 30–50% sucrose in many regular dry mixes without affecting the amount of sweetness. It is preferred over sucrose in many fruit-based drinks because of its cooling and flavor-enhancing properties. It can also reduce excessive sweetness and provide better shelf life in certain products. However, the hygroscopicity of fructose often results in caking during storage, especially in humid conditions.

Caking or lumping problems can also occur when the powder is dispersed into the liquid during preparation of the final product. This phenomenon is most often the result of the fine particle sizes of the powders. Increasing the proportion of sucrose or the crystal size in the formulation can help to eliminate this problem. Agglomeration or cocrystallization of the sweetener with other ingredients can also aid dispersion of the dry mix into the liquid.

ALCOHOLIC BEVERAGES

Fermentation of sweeteners is an important step in the production of alcoholic beverages such as wine, beer, and champagne. Maltose generated during the malting of barley is fermented by yeast to alcohol. The simple sugars present in grapes are fermented by yeasts to make wine. Often, naturally present carbohydrate-based sweeteners are adequate for the completion of fermentation to the desired alcohol content. If the fermentation is incomplete, sugars such as sucrose are added for the yeasts to ferment. Sucrose is also added to sweeten some fermented wines. Special alcoholic products made with honey are available.

Champagne production depends on a double fermentation process in which sucrose is added to initiate and complete the second fermentation step.

Liqueurs and cordials also depend on the addition of sweeteners for their sweetness, flavor, and body. Sucrose is often the sweetener of choice for white or colorless liqueurs since it does not undergo Maillard browning and therefore does not cause off-colors. High-quality low-ash sweeteners are also desired because, in some cases, alcohol can react with the carbohydrate-based sweetener, causing the precipitation of polysaccharides or salts.

Large quantities of carbohydrates from barley malt, corn and rice grits, and starch are used in making beer and ale. The corn and rice may be added during barley malting to provide an additional source of carbohydrate to be converted to sugars and fermented by yeasts. Some brewers use more refined sweeteners such as corn syrups, malt syrups, and dextrose for this purpose.

Regular-DE (42–62) and special brewers' syrups are preferred because they have several advantages over other adjuncts. They increase solubility, produce high-gravity brews, increase brew-house efficiency and output, give higher purity, improve filtration rates, and reduce haze problems.

Dairy-Based Foods

Carbohydrate-based sweeteners are essential ingredients in several dairy-based foods (e.g., ice cream and frozen desserts). Although lactose is naturally present in these foods, it has very low sweetening power and poor solubility. Therefore, in some food applications, lactose is most often used not as a sweetener but as a filler to provide bulk or as a browning agent because it undergoes Maillard browning. However, since it is naturally present in milk (at about 4.8% in whole milk), the characteristics it has as a simple sugar must be considered. Lactose crystallization, for example, is sometimes a problem in dairy-based foods such as ice cream, where it can cause sandiness in the final product.

EFFECTS OF SWEETENERS

Several characteristics important in quality dairy foods can be attributed to the addition of carbohydrate-based sweeteners.

Freezing point depression. The addition of a carbohydrate-based sweetener lowers the freezing point of the product. Therefore, at any given temperature, a frozen product containing a sweetener, such as sugar or HFCS, is less solid than a comparable product without sweetener. In general, the higher the DE of the sweetener, the more the freezing point is lowered. For sherbets, ices, and soft-serve products, glucose syrups or HFCSs are used so that the product is soft yet still frozen. However, if sweeteners are used at a level that makes the freezing point too low, the final product may not freeze or may not be sufficiently solid.

Lower-DE corn syrups, which allow the freezing point to be somewhat higher, are used to give freeze-thaw stability to products typically stored in supermarket and home freezers. The repeated freeze-thaw cycles that occur in these freezers result in ice crystal growth and have a deleterious effect on product quality. A more detailed explanation of the colligative properties of sweeteners is given in Chapter 5.

The "meltdown" of a frozen dessert is also affected by sweeteners. The meltdown test for a frozen product is one measure of its quality. The product should melt easily at room temperature and lose its shape, and the melted part should be smooth and homogenous, not separated or in curds. A major factor that affects meltdown is the level of solids in the product. Because sweeteners are a substantial part of the total solids in frozen dessert, they can affect the meltdown and thus the quality of the product.

Texture. Sweeteners add body to dairy-based foods. The higher molecular weight components of the sweetener control the size of the sugar crystals produced. Too many large crystals cause iciness or coarseness in frozen dessert. Ice crystals can also form if high molecular weight fractions of sugars, such as those in corn syrup, are not present to prevent crystal formation.

Syrups can increase the viscosity of dips and yogurts while still yielding a clean, sweet flavor. Lower-DE products are especially effective. If low-DE products are used in frozen desserts, the final product can become chewy and have poor meltdown characteristics. Therefore, lower levels of low-DE sweeteners are used in these products to achieve the desired texture effects while still improving the smoothness and body of the final product. Higher-DE corn syrups and corn syrup solids help to create firmer textures with greater body.

Sweetness. The desired sweetness level depends on the dairy-based product being formulated. Fruit-flavored yogurts, in general, have more added sweeteners than plain yogurt formulations. Sweeteners with characteristic flavors such as molasses and honey are also used to create distinctive products.

ICE CREAM AND FROZEN DESSERTS

Since ice cream is by far the largest frozen dessert category, the general characteristics for making ice creams are discussed here. Many of these principles also apply to the making of related products such as frozen yogurt, mousses, ice milks, sherbets, and ices.

Ice cream production starts with the making of an ice cream mix comprised of nonfat milk solids, fat, sweeteners, and often a stabilizer and an emulsifier. The dry ingredients are blended and then mixed with the liquid ingredients. The sweetening system can use sucrose as the sole ingredient or contain additional sweeteners such as corn syrups, molasses, honey, refiner's syrups, fructose, brown sugar, or maple sugar (Table 8-2). Total sweetener levels range from 12 to 18%, with most products in the 14–16% range. Ices and sherbets have a higher percentage of corn syrups (about 4%) as a part of their sweetening system.

The next step is pasteurization and homogenization of the mix. The processed mix is then frozen while being simultaneously whipped to incorporate air. The volume of ice cream obtained in excess of the initial volume of mix is the "overrun," which is a factor in determining the product's final body and texture. Products with too high an overrun are too fluffy in texture, whereas products with too low an overrun are soggy and too dense. Various products have different desired overruns. In general, the higher the total solids level of the formulation, the higher the overrun needed to produce an acceptable product.

TABLE 8-2. Sweetener Content (%) of Typical Dairy Products and Desserts

Product	Sucrose	Corn Syrup, 42 or 62 DE[a]	HFCS[b]
Ice cream, 12 % fat	12.0	5.0	
		9.0	9.0
	6.0	6.0	6.0
Soft-serve, low fat	11.0	7.6	
Sherbet	18.0	10.0	
		14.0	14.0
Frozen yogurt	11.0	6.0	
		8.5	8.5
Cream pie filling			
Chocolate	18.4	6.1	
Coconut	18.4	6.1	
Coffee whitener		45.0[c]	

[a] Dextrose equivalent.
[b] High-fructose corn syrup.
[c] 36-DE corn syrup.

Overrun is influenced by several factors such as ingredient selection, formulation, and processing steps. Sweetener choice and level play an important role in determining the overrun. In general, overrun is reduced when corn syrup solids are used. This fact is useful in formulating ice creams, as it allows the amount of overrun to be controlled.

After the mix has been whipped and frozen, it is either molded or packaged and then further hardened by storage at a cold temperature. Final product defects such as weak body or sandiness can be related to the sweetening system of the formulation. Ice creams of high quality often contain only sucrose as sweetener. The sucrose content of sherbets and ices is about twice that of ice cream. However, sucrose alone is not a good sweetening system for sherbets and ices because it can crystallize at the product surface. Dextrose, which helps to inhibit sucrose crystallization in these products, is added in the form of corn syrup solids or corn syrup at about one-fourth to one-third the amount of sucrose present in the formula. This helps to ensure that the sherbet or ice does not become too hard and crumbly by further lowering its freezing point.

Sandiness in frozen dairy products is the result of lactose crystallization. This defect can be controlled by a decrease in lactose concentration or by the addition of a dextrose-containing sweetener such as corn syrup solids or HFCS. In addition to reducing sandiness, glucose syrups help to control the body and texture of the final product as well as the sweetness level. Higher-DE syrups also are helpful in creating smoother end products, but if high-DE syrups are used at too high a level (as much as 5% too high), they can cause chewiness.

YOGURT

Yogurt is a fermented product, which thus depends on the presence of a fermentable carbohydrate. Lactose, naturally occurring in milk, provides the raw material that the fermenting microorganisms convert to acid. The production of acid causes the milk proteins to denature and coagulate, causing the system to gel and set. Additional sweeteners such as HFCS, corn syrups, and dextrose may be added either to the initial formulation or during the addition of fruit to help develop the final flavor of the product (Fig. 8-2). While sucrose is the most widely used sweetener, special yogurts that use honey, brown sugars, or molasses are also produced.

Fig. 8-2. Yogurt contains lactose and added sweeteners. (Courtesy A. E. Staley Manufacturing Co.)

Soups, Gravies, and Meats

Soups and gravies. A few canned soup products use sweeteners, either HFCS or sucrose. Dry soup mixes employ corn syrup or corn syrup solids (CSS) plus some sucrose. Maltodextrins are used as carriers and bulking agents in a variety of dry mixes, including beef- and chicken-flavored soups, gravies, and sauces. They offer bland flavor, low hygroscopicity, rapid dispersibility, and high solubility. Products like beef or chicken nuggets can contain CSS or maltodextrins plus sucrose.

Canned foods and preserved meats. In high-temperature processing, sucrose is often preferred over many other sweeteners because it is a nonreducing sugar and does not produce dark colors as easily as reducing sugars do.

In canned and preserved meats, however, either crystalline fructose or dextrose are often preferred over sucrose. They do not produce undesirable microbial slime, as sucrose sometimes does, and they improve colors and flavors. Some meat processors are switching to liquid dextrose (sometimes referred to as dextrose syrup) because it is less expensive and easier to use.

Some corn syrups (low-DE products), CSS, and maltodextrins are also used in processed meats for their fat- and water-binding properties. In addition, lactose is employed to mask certain salts and flavors and reduce sweetness, while increasing solids in brine mixtures.

Troubleshooting

JAMS/JELLIES/PRESERVES

Symptoms	Causes	Changes to Make
Gritty/sandy texture	Sugar crystallization	Reduce soluble solids to <78%. Increase ratio of glucose-containing syrups to sucrose. Substitute maltose syrup for dextrose if dextrose content exceeds 35%.
Spoilage	Water activity too high	Increase sweetener level.
Thin consistency or lack of structure	Low gel strength	If high-methoxy pectin used, increase sugar. If low-methoxy pectin used, check pH and/or add more pectin if necessary.
Texture too hard and/or brittle structure	Gel strength too high	If high-methoxy pectin used, decrease sugar. If low-methoxy pectin used, check pH and/or decrease pectin if necessary.
Gel breakdown	Sucrose inversion	Decrease boiling time. Increase cooling time. Check pH.

BEVERAGES, LIQUID

Symptoms	Causes	Changes to Make
Lacks body and/or mouthfeel	Low viscosity or solids content	Add lower-DE syrups and/or maltodextrins to add mouthfeel without increasing sweetness.
Off flavor/odor	Microbial or yeast growth	Check quality of incoming ingredients.
Change in sweetness level	Sucrose inversion or sweetener breakdown	Check pH. Decrease storage temperature and/or time.

BEVERAGES, POWDERED MIXES

Symptoms	Causes	Changes to Make
Caking	Excessive moisture in product	Increase sweetener particle size. Decrease hygroscopic sweetener levels (i.e., fructose, dextrose). Use anticaking agents. Store ingredients in dry environment.
Powder mix not homogeneous	Settling of large particles	Decrease sweetener particle size.
Poor dispersion into solution	Poor particle wetting	Increase sweetener particle size. Agglomerate product.

BEVERAGES, ALCOHOLIC

Symptoms	Causes	Changes to Make
Low alcohol level	Incomplete/insufficient fermentation	Increase fermentable carbohydrates by increasing levels of sweeteners such as maltose, HFCS, or sucrose.
Color too dark	Excessive Maillard browning	Decrease reducing sugar level or increase sucrose level (if possible). Check pH for sucrose inversion.
Sedimentation	Precipitation of sweetener with salts or other polysaccharides	Check quality of incoming sweeteners (low ash desired).

FROZEN DESSERTS

Symptoms	Causes	Changes to Make
Too hard	Freezing point too high	Increase sweetener level.
Too soft, icy, weak body	Freezing point too low	Decrease sweetener level.
Lacks body	Low solids content	Increase sweetener level. Add syrups such as HFCS.
Chewy or gummy texture	Excessive moisture binding by ingredients	Decrease corn syrup solids and/or low-DE sweeteners. Adjust stabilizer level.
Sandy or gritty texture	Excessive lactose crystallization	Decrease lactose content.
	Sucrose crystallization	Add dextrose in the form of corn syrups or corn syrup solids.
Texture too dense	Low overrun or too little air incorporated	Decrease corn syrups.
Fluffy texture	High overrun or too much air incorporated	Increase corn syrups.

Special Topics

Sweetener Selection in Product Development

During the development of a new product, nutritional concerns, labeling regulations, and consumer preferences are all considered, as well as the functional attributes of the ingredients and how the ingredients interact during processing and in the final product. When formulations for sweetened products are developed, several factors regarding the choice of sweetener must be considered. Obviously, a choice among dry granular products (e.g., sucrose, dextrose, fructose, or corn syrups solids) must be made for the production of dry-mix foods such as beverage mixes, dried sauces, or cake and muffin mixes. For ready-to-eat, packaged foods, the list of potential sweeteners expands to include syrups such as corn syrups, high-fructose corn syrups (HFCSs), malt syrups, molasses, honey, refiners syrups and the like. Since sweeteners influence several factors, including food preservation, browning, texture, flavor, and fermentability, each of these factors must be considered when developing a sweetened food product.

Flavor characteristics. What final product flavor is desired by the consumer? Distinctive flavors are inherent in sweeteners such as molasses, honey, malt, and refiners syrups. These sweeteners may enhance the characteristics of the final product or may overwhelm the delicate flavors desired in the formulated food. Clean-tasting sweeteners such as sucrose or corn syrups may be necessary in delicately flavored products.

Texture. What textural characteristics are necessary for the product? Chewiness, crispness, mouthfeel, and body are all influenced by the type of sweetener chosen. A sweetener may affect more than one aspect of texture, however. For instance, corn syrups can help increase body and mouthfeel but may also increase the chewiness of the final product.

Product storage. Will the product remain stable over its expected shelf life? Sweeteners reduce the water activity, which makes the environment less supportive to microorganisms. They also bind water, thereby helping the product to retain moistness and softness. However, high-DE sugars can pick up moisture over time, causing dry mixes to cake or candies to have a sticky surface film.

Turbinado sugar—A raw sugar that has been partially processed, removing some of the surface molasses. It is blond, has a mild flavor, and is often used in tea.

Product identity. What should the final product look like? Should it have a shiny glaze or a matte, frosted appearance? Should it be translucent or opaque, colored or clear? As each of these questions is addressed, the answer directly affects sweetener choice. Syrups or high-DE products can give a glazed appearance, while sucrose may be necessary to create a frosted look. If browning is desired, sugars other than sucrose are needed unless acids or higher temperatures are used to induce caramelization of sucrose or cause inversion to glucose and fructose, which can undergo Maillard browning.

Processing requirements. Is it necessary to use dry, granular products, or can liquids be pumped into the production system? Is a dry sweetener necessary to disperse or carry ingredients such as vitamins or flavors? These questions are important to consider in choosing between syrups and granular sweeteners.

Nutritional concerns. Lactose intolerance is a condition in which a person is unable to digest lactose because his or her body lacks the enzyme β-galactosidase, which cleaves lactose into its two component sugars, galactose and glucose. Such individuals suffer from bloating and flatulence if they consume lactose. Therefore, manufacturers of food product sometimes avoid lactose as an ingredient in foods. Other individuals affected by carbohydrate-based sweeteners are those with diabetes (see below). Often, fructose is chosen as a sweetener rather than sucrose because it can be digested without the presence of insulin.

TABLE 9-1. Quantities of Sweeteners Used in Major Food Applications[a]

Food Category	Quantity	
	Billion pounds, dry basis	Percentage
Dairy products	2.966	9.5
Preserved fruits and vegetables	2.882	9.3
Grain mill products	3.080	9.9
Bakery products	3.626	11.6
Confectionery	4.521	14.5
Beverages	13.064	42.0
Miscellaneous foods	0.996	3.2

[a] Data from Census of Manufactures (Dept. of Commerce, Bureau of the Census) for 1992.

Cost and availability. Large quantities of sweeteners are used each year in the food industry (Table 9-1). U. S. per capita sweetener consumption for 1985–1996 is shown in Table 9-2. For any given product, the choice of a specific sweetener may depend upon the relative costs and availability of alternative sweeteners. As market prices fluctuate, substitution of one sweetener for another (e.g., corn syrups for sucrose, or vice versa) may provide a cost savings. Although specialty sweeteners such as *turbinado sugar* or raw sugars may be desired for a product, market conditions may make them cost prohibitive. Thus, within an industry segment, the proportions of sweeteners used may fluctuate over time.

Combinations. The above factors can be thought of individually, but their interactions must also be considered. For example, corn syrups may reduce the cost of the final product, but (depending on other ingredients and production factors) they may also cause a glazed appearance or chewiness. Each factor should be considered with the others in formulating the desired product.

TABLE 9-2. U.S. Per Capita Sweetener Consumption[a] (pounds)

Year	Refined Sugar	HFCS[b]	Glucose	Dextrose	Total Corn Syrups	Honey and Edible Syrups	Total Caloric Sweeteners
1985	63.2	44.6	15.9	3.5	63.9	1.5	128.6
1986	60.8	45.1	16.0	3.5	64.6	1.6	127.0
1987	63.1	47.1	16.2	3.6	66.8	1.7	131.6
1988	62.6	48.3	16.4	3.6	68.3	1.5	132.4
1989	62.8	47.5	16.7	3.7	68.0	1.6	132.4
1990	64.8	49.2	17.4	3.8	70.4	1.6	136.8
1991	64.4	50.0	18.2	3.8	72.0	1.6	138.0
1992	64.4	51.6	19.0	3.8	74.3	1.6	140.3
1993	64.6	54.4	19.6	3.8	77.8	1.6	144.0
1994	65.8	56.4	20.0	3.9	80.2	1.5	147.5
1995	66.2	58.4	20.3	4.0	82.6	1.5	150.3
1996[c]	66.9	59.8	20.4	4.0	84.2	1.5	152.6

[a] Source: USDA Economic Research Service.
[b] High-fructose corn syrup.
[c] Estimate.

Dental Caries

Many foods are cariogenic; that is, they promote cavities in teeth. Dental caries is a complex phenomenon influenced by several factors such as the microbial environment of an individual's mouth, the structure of the teeth, and the individual's oral hygiene. Essentially, microorganisms in the mouth ferment the carbohydrates that are present. The end products of fermentation include polysaccharides, which bind to the teeth, thus forming plaque, which can eventually lead to the formation of cavities. Sweeteners, therefore, play a role in dental caries because they are fermentable carbohydrates that microorganisms are able to use. Carbohydrate-based sweeteners alone are not responsible for dental caries, but they do influence the occurrence of caries. The incidence of dental caries has declined significantly with the addition of fluoride to water supplies as well as to hygiene products such as toothpastes.

Diabetes

When carbohydrates are eaten, a series of hydrolysis steps converts them into their component simple sugars. These sugars then undergo a series of biochemical conversions via metabolic pathways. Some are metabolized directly, while most are converted to D-glucose. Glucose, the bodys source of energy, is easily absorbed into the bloodstream. After a meal, a large increase in the blood glucose level occurs. This is

TABLE 9-3. Typical Glycemic Index (GI) Values of Selected Foods and Sweeteners

	Glycemic Index
Food	
Potatoes	95
White rice	95
Chocolate bar	70
Cookies	70
Banana	60
Jam	55
Ice cream	50
Oatmeal	50
Green vegetables	<15
Sweetener	
Maltose	110
Glucose	100
Honey	90
Sucrose	90
Fructose	20

Glycemic index—A measure of how much a food increases the glucose level in blood after it is digested.

measured by the *glycemic index*. Different foods have different glycemic index values (Table 9-3).

In response to the increase in blood glucose, the pancreas secretes the hormone insulin, which controls blood glucose levels. Individuals who are diabetic either are unable to produce insulin or secrete insulin that is not able to function properly. Therefore, their "blood sugar" (blood glucose level) remains high, which can lead to shock and even death. Through the injection of insulin, many people with diabetes are able to control their blood glucose levels. In less severe cases, these levels can also be controlled through the diet by monitoring the consumption of foods with a high glycemic index and regulating the amounts eaten. These foods can also be combined with other foods that have a low glycemic index. Fructose is a unique sugar in that it does not depend on the presence of insulin for its metabolism. It has a very low glycemic index and is often used in formulating diabetic foods.

Regulatory Status and Nutritional Labeling

The U.S. Food and Drug Administration considers many foods and food ingredients in general use before 1958 to be generally recognized as safe (GRAS). This category was created by the Food Additives Amendment of 1958, which modified the U.S. Food, Drug and Cosmetic Act of 1938. Later rulings confirmed and extended the category. Sucrose, invert sugar, corn syrups, dextrose, lactose, fructose, HFCS, and malt syrups are GRAS.

In 1990, Congress passed the Nutritional Labeling and Education Act (NLEA), which was followed in 1993 by regulations from the Food and Drug Administration (FDA) for the labeling of foods. These regulations clarified several areas involving carbohydrate-based sweeteners. First, sugars became a mandatory nutrient to be declared on the food label. Previously, their declaration was not required un-

---

Box 9-1. Regulations for Nutrient Content Claims for Sugars, as of January 1993

Sugar free: must contain less than 0.5 g of sugars per reference amount of food. Cannot apply to an ingredient that is a sugar or generally is understood to contain sugars.

Low-sugar: not defined. No recommended intake given.

Reduced or less sugar: must contain at least 25% less sugar per reference amount than an appropriate reference food contains.

No added sugars (or Without added sugars): claim may be made if no sugar or sugar-containing ingredient is added during processing

less a nutrient claim was made involving sugars, for example a low-sugar claim. Before the NLEA, no definition of "sugar" existed, and manufacturers may have declared only the sucrose concentration of the food. The NLEA defined sugars as the sum of all the free mono- and disaccharides present in the food. This definition had the most impact on foods with naturally occurring mono- and disaccharides, which were not previously declared. The most notable example is milk, for which the lactose content is now required on the label under the sugars declaration (Fig. 9-1).

Also clearly defined were nutrient content claims for sugars (sugar-free, low-sugar, reduced/less sugar), which are explained in Box 9-1. The most notable change here is that, in sugar-free claims, no ingredients in the food can contain sugar of any kind. For example, a claim such as "sugar free" peaches would not be allowed because peaches contain fructose, which now falls under the definition for sugar and is present at a level greater than 0.5 g per reference amount.

Nutrition Facts

Serving Size 1 cup (240 mL)
Servings Per Container 1

Amount Per Serving

Calories 130	Calories from Fat 45
	% Daily Value*
Total Fat 5g	8%
Saturated Fat 3g	15%
Cholesterol 20mg	7%
Sodium 125mg	5%
Total Carbohydrate 12g	4%
Dietary Fiber 0g	0%
Sugars 12g	
Protein 9g	16%

Vitamin A 10% • Vitamin C 2% • Calcium 30%

Iron 0% • Vitamin D 25%

*Percent Daily Values are based on a 2,000 calorie diet. Your daily values may be higher or lower depending on your calorie needs:

		Calories:	2,000	2,500
Total Fat	Less than		65g	80g
Sat. Fat	Less than		20g	25g
Cholesterol	Less than		300mg	300mg
Sodium	Less than		2400mg	2400mg
Total Carbohydrate			300g	375g
Dietary Fiber			25g	30g

Calories Per Gram:
Fat 9 • Carbohydrate 4 • Protein 4

Fig. 9-1. Typical nutrition label for reduced-fat milk. Because lactose is naturally present in milk, the label must show that the milk contains sugar (see shaded area).

APPENDIX A.

Characteristics of Selected Sugars

Sugar	Formula	Relative Sweetness	MW[a]	Source	Descriptions
Monosaccharides					
Fructose	$C_6H_{12}O_6$	1.4–1.6	180	Natural in fruits and honey Isomerized from glucose	Can be digested without insulin
Glucose (dextrose)	$C_6H_{12}O_6$	0.7–0.8	180	From hydrolysis of starch	Often sold as a monohydrate. Check supplier specifications.
Disaccharides					
Lactose	$C_{12}H_{22}O_{11}$	0.4	342	Galactose + glucose	Insoluble sugar derived from processing of dairy foods. Very low sweetness
Maltose	$C_{12}H_{22}O_{11}$	0.45	342	Glucose + glucose	Often used in syrup form as a source of fermentable carbohydrates
Sucrose	$C_{12}H_{22}O_{11}$	1.00	342	Glucose + fructose	Clean-tasting. No Maillard browning

[a] Molecular weight.

APPENDIX B.

Characteristics of Additional Carbohydrate-Based Products

TABLE B-1. Properties of Granulated Sugars

Property	Coarse	Sanding	Extra Fine	Fruit	Bakers' Special	Powdered 6X	Powdered 10X
Percent on screen							
U.S. 12	0–2	...	...	...	...	...	...
U.S. 16	45–65	...	...	...	...	...	...
U.S. 20	20–35	2–10	0–5	...	...	...	...
U.S. 30	...	40–70	2–20	...	...	...	...
U.S. 40	...	30–40	10–45	0–7	...	...	...
U.S. 50	...	1–8	5–35	20–50	0–5	...	...
U.S. 70	...	...	...	30–70	10–30	...	...
U.S. 100	...	...	10–40	10–30	30–60	0–2	0–1
Percent through screen							
U.S. 100	...	...	0–8	0–10	10–30	...	...
U.S. 140	...	...	...	...	5–20	...	...
U.S. 200	...	...	...	...	...	88–100	94–100
Color, ICU	20–35	20–35	25–50	25–50	25–50	25–50	25–50
Ash, % (max)	0.015	0.015	0.02	0.03	0.03	0.03	0.03
Moisture, % (max)	0.04	0.04	0.05	0.05	0.05	0.5	0.5
Starch, %	...	...	...	...	...	2.5–3.5	2.5–3.5

TABLE B-2. Properties of Liquid Sugar

Property	Liquid Sucrose	Amber Sucrose	Liquid Invert	Total Invert
Solids, %	67.0–67.9	67.0–67.7	76–77	71.5–73.5
Invert, %	≤0.35	≤0.4	45–55	≥93
Sucrose, %	≥99.5	≥99.4	55–45	≤7
Glucose, %	...	...	23–28	≥46.5
Fructose, %	...	...	23–28	≥46.5
Color, ICU	≤35	≤200	≤35	≤40
Ash, %	≤0.04	≤0.15	≤0.06	≤0.09
pH	6.7–8.5	6.5–8.5	4.5–5.5	3.5–4.5
Solids, lb/gal	7.42–7.55	7.42–7.52	8.75–8.91	8.05–8.35
Total weight, lb/gal	11.08–11.12	11.08–11.11	11.52–11.57	11.25–11.36

TABLE B-3. Properties of Brown Sugars

Property	Soft Brown		Coated Brown		Free–Flowing	
	Light	Dark	Light	Dark	Granulated	Powdered
Sucrose, %	85–93	85–93	90–96	90–96	91–96	91–96
Invert, %	1.5–4.5	1.5–5	2–5	2–5	2–6	2–6
Ash, %	1–2	1–2.5	0.3–1	0.3–1	1–2	1–2
Organic nonsugars, %	2–4.5	2–4.5	1–3	1–3	0.5–1	0.5–1
Moisture, %	2–3.5	2–3.5	1–2.5	1–2.5	0.4–0.9	0.4–0.9
Color, ICU	3,000–6,000	7,000–11,000	3,000–6,000	7,000–11,000	6,000–8,000	6,000–8,000
Color, reflectance	40–60	25–35	...	...	...	...
Granulation On U.S. 16, %	...	...	...	...	≤6	0
Through U.S. 50, %	...	...	...	...	≤8	≥86
Through U.S. 100, %	...	...	...	...	...	≤65

TABLE B-4. Characteristics of Typical Maltodextrins[a]

Type (DE)[b]	Percent Solids	Actual DE	pH	Carbohydrate Composition[c]			
				DP1	DP2	DP3	Higher DP
5-W	95.0	5–8	4.5–5.0	<0.5	1	1.5	97+
5-C	94–95	4–7	4.0–5.0	<1.0	<1.0	1.0	97+
10-W	95.0	9–13	4.5–5.0	0.5	2	3	94+
10-C	94–95	9–12	4.0–5.0	1	2	3	93+
15-W	95.0	14–18	4.5–5.0	2	3	4	90+
15-C	94–95	13–17	4.0–5.0	2	3	4	90+
18-C	94–95	16.5–19.5	4.0–5.0				

[a] A more detailed listing of the available maltodextrins and their properties was reported by Alexander (1).

[b] DE = dextrose equivalent, W = waxy starch, C = common starch.

[c] DP = degree of polymerization. DP1 = glucose, DP2 = maltose, etc.

TABLE B-5. Characteristics of Typical Corn Syrups and Corn Syrup Solids

Type (DE)[a]	Percent Solids	Actual DE	pH	Carbohydrate Composition[b]				Relative Sweetness[c]
				DP1	DP2	DP3	Higher DP	
Corn syrups								
24	78	24–26	4.5–5.0	5	6	11	78	...
28	78–80	27–29	4.5–5.0	6–8	7–15	8–12	70–75	0.35
36	80	35–37	4.5–5.0	13–14	11–12	10	64–66	0.40
42	80–82	41–43	4.5–5.0	19	13–14	12	55	0.50
54	81	52–54	4.5–5.0	28–30	17–18	13	39–42	0.55
62	81–84	61–63	4.5–5.0	36–37	32	12–14	19	0.65
66	82–84	65–66	4.0–4.5	40	35	8	11	...
70	82	69–70	4.5–5.0	42	37	5	16	...
80	80	80	4.0–5.0	56	...	...	...	...
Corn syrup solids								
20	95	18–22	4.0–5.0	1–2	6–7	8–9	83–85	...
24	93–95	23–27	4.0–5.0	2–6	5–8	6–10	76–84	...
30	95	28–32	4.5–5.0	3	9	9	79	...
36	95–96	34–38	4.0–5.0	13–14	11–12	10	64–66	...
42	95–96	40–44	4.0–5.0	19	12–14	11–12	55–58	...
36HM[d]	95	34–38	4.5–5.0	6	29	11	54	...
42HM	95	40–44	4.5–5.0	7	39	17	37	...
48HM	97	48	4.0–4.5	4	65	15	16	...

[a] Dextrose equivalent.
[b] DP = degree of polymerization. DP1 = glucose, DP2 = maltose, etc.
[c] Sweetness relative to sucrose at 1.0.
[d] HM = high maltose.

TABLE B-6. Characteristics of Typical Dextrose Syrups and High-Maltose Corn Syrups

Type	Percent Solids	Actual DE[a]	pH	Carbohydrate Composition[b]			
				DP1	DP2	DP3	Higher DP
Dextrose syrups (DE)							
95	71	95+	4.5	95	3	0.5	1.5
99	70	99.5	4.0	99	0.07	...	0.3
	81	99.5	4.0	99	0.07	...	0.3
High-maltose corn syrups (% maltose)							
32	80	43	5.0	10	32	...	...
40	81	50	5.0	12	40	...	...
45	80	42	4.5	8	45	15	32
51	81	48	4.5	8	51	24	17

[a] Dextrose equivalent.
[b] DP = degree of polymerization. DP1 = glucose, DP2 = maltose, etc.

TABLE B-7. Vitamin and Mineral Content of Honey[a,b]

Vitamins	mg/100 g	Minerals	mg/100 g
Thiamin	0.006	Calcium	4.4–9.2
Riboflavin	0.06	Copper	0.003–0.10
Niacin	0.36	Iron	0.06–1.5
Pantothenic acid	0.11	Magnesium	1.2–3.5
Pyridoxine (B-6)	0.32	Manganese	0.02–0.4
Ascorbic acid	2.2–2.4	Phosphorus	1.9–6.3
		Potassium	13.2–168
		Sodium	0.0–7.6
		Zinc	0.03–0.4

[a] pH 3.9.
[b] Data from (2).

References

1. Alexander, E. 1982. Polyols: Chemistry and applications. In: *Food Carbohydrates*. D. R. Lineback and G. E. Inglett, eds. AVI Publishing Co., Westport, CT.
2. Anonymous. 1988. Honey: From Nature's Food Industry. National Honey Board, Longmont, CO.

APPENDIX C.

Specific Gravity, Degrees Brix, and Degrees Baumé of Sugar Solutions

Degrees Brix or % by Weight of Sucrose	Specific Gravity at 20/20°	Specific Gravity at 20/4°	Degrees Baumé (Modulus 145)	Degrees Brix or % by Weight of Sucrose	Specific Gravity at 20/20°	Specific Gravity at 20/4°	Degrees Baumé (Modulus 145)
0.0	1.00000	0.998234	0.00	9.0	1.03586	1.034029	5.02
0.2	1.00078	0.999010	0.11	9.2	1.03668	1.034850	5.13
0.4	1.00155	0.999786	0.22	9.4	1.03750	1.035671	5.24
0.6	1.00233	1.000563	0.34	9.6	1.03833	1.036494	5.35
0.8	1.00311	1.001342	0.45	9.8	1.03915	1.037318	5.46
1.0	1.00389	1.002120	0.56	10.0	1.03998	1.038143	5.57
1.2	1.00467	1.002897	0.67	10.2	1.04081	1.038970	5.68
1.4	1.00545	1.003675	0.79	10.4	1.04164	1.039797	5.80
1.6	1.00623	1.004453	0.90	10.6	1.04247	1.040626	5.91
1.8	1.00701	1.005234	1.01	10.8	1.04330	1.041456	6.02
2.0	1.00779	1.006015	1.12	11.0	1.04413	1.042288	6.13
2.2	1.00858	1.006796	1.23	11.2	1.04497	1.043121	6.24
2.4	1.00936	1.007580	1.34	11.4	1.04580	1.043954	6.35
2.6	1.01015	1.008363	1.46	11.6	1.04664	1.044788	6.46
2.8	1.01093	1.009148	1.57	11.8	1.04747	1.045625	6.57
3.0	1.01172	1.009934	1.68	12.0	1.04831	1.046462	6.68
3.2	1.01251	1.010721	1.79	12.2	1.04915	1.047300	6.79
3.4	1.01330	1.011510	1.90	12.4	1.04999	1.048140	6.90
3.6	1.01409	1.012298	2.02	12.6	1.05084	1.048980	7.02
3.8	1.01488	1.013089	2.13	12.8	1.05168	1.049822	7.13
4.0	1.01567	1.013881	2.24	13.0	1.05252	1.050665	7.24
4.2	1.01647	1.014673	2.35	13.2	1.05337	1.051510	7.35
4.4	1.01726	1.015467	2.46	13.4	1.05422	1.052356	7.46
4.6	1.01806	1.016261	2.57	13.6	1.05506	1.053202	7.57
4.8	1.01886	1.017058	2.68	13.8	1.05591	1.054050	7.68
5.0	1.01965	1.017854	2.79	14.0	1.05677	1.054900	7.79
5.2	1.02045	1.018652	2.91	14.2	1.05762	1.055751	7.90
5.4	1.02125	1.019451	3.02	14.4	1.05847	1.056602	8.01
5.6	1.02206	1.020251	3.13	14.6	1.05933	1.057455	8.12
5.8	1.02286	1.021053	3.24	14.8	1.06018	1.058310	8.23
6.0	1.02366	1.021855	3.35	15.0	1.06104	1.059165	8.34
6.2	1.02447	1.022659	3.46	15.2	1.06190	1.060022	8.45
6.4	1.02527	1.023463	3.57	15.4	1.06276	1.060880	8.56
6.6	1.02608	1.024270	3.69	15.6	1.06362	1.061738	8.67
6.8	1.02689	1.025077	3.80	15.8	1.06448	1.062598	8.78
7.0	1.02770	1.025885	3.91	16.0	1.06534	1.063460	8.89
7.2	1.02851	1.026694	4.02	16.2	1.06621	1.064324	9.00
7.4	1.02932	1.027504	4.13	16.4	1.06707	1.065188	9.11
7.6	1.03013	1.028316	4.24	16.6	1.06794	1.066054	9.22
7.8	1.03095	1.029128	4.35	16.8	1.06881	1.066921	9.33
8.0	1.03176	1.029942	4.46	17.0	1.06968	1.067789	9.45
8.2	1.03258	1.030757	4.58	17.2	1.07055	1.068658	9.56
8.4	1.03340	1.031573	4.69	17.4	1.07142	1.069529	9.67
8.6	1.03422	1.032391	4.80	17.6	1.07229	1.070400	9.78
8.8	1.03504	1.033209	4.91	17.8	1.07317	1.071273	9.89

(continued)

Specific Gravity, Degrees Brix, and Degrees Baumé of Sugar Solutions, continued

Degrees Brix or % by Weight of Sucrose	Specific Gravity at 20/20°	Specific Gravity at 20/4°	Degrees Baumé (Modulus 145)	Degrees Brix or % by Weight of Sucrose	Specific Gravity at 20/20°	Specific Gravity at 20/4°	Degrees Baumé (Modulus 145)
18.0	1.07404	1.072147	10.00	27.0	1.11480	1.112828	14.93
18.2	1.07492	1.073023	10.11	27.2	1.11573	1.113763	15.04
18.4	1.07580	1.073900	10.22	27.4	1.11667	1.114697	15.15
18.6	1.07668	1.074777	10.33	27.6	1.11761	1.115635	15.26
18.8	1.07756	1.075657	10.44	27.8	1.11855	1.116572	15.37
19.0	1.07844	1.076537	10.55	28.0	1.11949	1.117512	15.48
19.2	1.07932	1.077419	10.66	28.2	1.12043	1.118453	15.59
19.4	1.08021	1.078302	10.77	28.4	1.12138	1.119395	15.69
19.6	1.08110	1.079187	10.88	28.6	1.12232	1.120339	15.80
19.8	1.08198	1.080072	10.99	28.8	1.12327	1.121284	15.91
20.0	1.08287	1.080959	11.10	29.0	1.12422	1.122231	16.02
20.2	1.08376	1.081848	11.21	29.2	1.12517	1.123179	16.13
20.4	1.08465	1.082737	11.32	29.4	1.12612	1.124128	16.24
20.6	1.08554	1.083628	11.43	29.6	1.12707	1.125076	16.35
20.8	1.08644	1.084520	11.54	29.8	1.12802	1.126030	16.46
21.0	1.08733	1.085414	11.65	30.0	1.12898	1.126984	16.57
21.2	1.08823	1.086309	11.76	30.2	1.12993	1.127939	16.67
21.4	1.08913	1.087205	11.87	30.4	1.13089	1.128896	16.78
21.6	1.09003	1.088101	11.98	30.6	1.13185	1.129853	16.89
21.8	1.09093	1.089000	12.09	30.8	1.13281	1.130812	17.00
22.0	1.09183	1.089900	12.20	31.0	1.13378	1.131773	17.11
22.2	1.09273	1.090802	12.31	31.2	1.13474	1.132735	17.22
22.4	1.09364	1.091704	12.42	31.4	1.13570	1.133698	17.33
22.6	1.09454	1.092607	12.52	31.6	1.13667	1.134663	17.43
22.8	1.09545	1.093513	12.63	31.8	1.13764	1.135628	17.54
23.0	1.09636	1.094420	12.74	32.0	1.13861	1.136596	17.65
23.2	1.09727	1.095328	12.85	32.2	1.13958	1.137565	17.76
23.4	1.09818	1.096236	12.96	32.4	1.14055	1.138534	17.87
23.6	1.09909	1.097147	13.07	32.6	1.14152	1.139506	17.98
23.8	1.10000	1.098058	13.18	32.8	1.14250	1.140479	18.08
24.0	1.10092	1.098971	13.29	33.0	1.14347	1.141453	18.19
24.2	1.10183	1.099886	13.40	33.2	1.14445	1.142429	18.30
24.4	1.10275	1.100802	13.51	33.4	1.14543	1.143405	18.41
24.6	1.10367	1.101718	13.62	33.6	1.14641	1.144384	18.52
24.8	1.10459	1.102637	13.73	33.8	1.14739	1.145363	18.63
25.0	1.10551	1.103557	13.84	34.0	1.14837	1.146345	18.73
25.2	1.10643	1.104478	13.95	34.2	1.14936	1.147328	18.84
25.4	1.10736	1.105400	14.06	34.4	1.15034	1.148313	18.95
25.6	1.10828	1.106324	14.17	34.6	1.15133	1.149298	19.06
25.8	1.10921	1.107248	14.28	34.8	1.15232	1.150286	19.17
26.0	1.11014	1.108175	14.39	35.0	1.15331	1.151275	19.28
26.2	1.11106	1.109103	14.49	35.2	1.15430	1.152265	19.38
26.4	1.11200	1.110033	14.60	35.4	1.15530	1.153256	19.49
26.6	1.11293	1.110963	14.71	35.6	1.15629	1.154249	19.60
26.8	1.11386	1.111895	14.82	35.8	1.15729	1.155242	19.71

(continued)

Specific Gravity, Degrees Brix, and Degrees Baumé of Sugar Solutions, continued

Degrees Brix or % by Weight of Sucrose	Specific Gravity at 20/20°	Specific Gravity at 20/4°	Degrees Baumé (Modulus 145)	Degrees Brix or % by Weight of Sucrose	Specific Gravity at 20/20°	Specific Gravity at 20/4°	Degrees Baumé (Modulus 145)
36.0	1.15828	1.156238	19.81	45.0	1.20467	1.202540	24.63
36.2	1.15928	1.157235	19.92	45.2	1.20573	1.203603	24.74
36.4	1.16028	1.158233	20.03	45.4	1.20680	1.204668	24.85
36.6	1.16128	1.159233	20.14	45.6	1.20787	1.205733	24.95
36.8	1.16228	1.160233	20.25	45.8	1.20894	1.206801	25.06
37.0	1.16329	1.161236	20.35	46.0	1.21001	1.207870	25.17
37.2	1.16430	1.162240	20.46	46.2	1.21108	1.208940	25.27
37.4	1.16530	1.163245	20.57	46.4	1.21215	1.210013	25.38
37.6	1.16631	1.164252	20.68	46.6	1.21323	1.211086	25.48
37.8	1.16732	1.165259	20.78	46.8	1.21431	1.212162	25.59
38.0	1.16833	1.166269	20.89	47.0	1.21538	1.213238	25.70
38.2	1.16934	1.167281	21.00	47.2	1.21646	1.214317	25.80
38.4	1.17036	1.168293	21.11	47.4	1.21755	1.215395	25.91
38.6	1.17138	1.168307	21.21	47.6	1.21863	1.216476	26.01
38.8	1.17239	1.170322	21.32	47.8	1.21971	1.217559	26.12
39.0	1.17341	1.171340	21.43	48.0	1.22080	1.218643	26.23
39.2	1.17443	1.172359	21.54	48.2	1.22189	1.219729	26.33
39.4	1.17545	1.173379	21.64	48.4	1.22298	1.220815	26.44
39.6	1.17648	1.174400	21.75	48.6	1.22406	1.221904	26.54
39.8	1.17750	1.175423	21.86	48.8	1.22516	1.222995	26.65
40.0	1.17853	1.176447	21.97	49.0	1.22625	1.224086	26.75
40.2	1.17956	1.177473	22.07	49.2	1.22735	1.225180	26.86
40.4	1.18058	1.178501	22.18	49.4	1.22844	1.226274	26.96
40.6	1.18162	1.179527	22.29	49.6	1.22954	1.227371	27.07
40.8	1.18265	1.180560	22.39	49.8	1.23064	1.228469	27.18
41.0	1.18368	1.181592	22.50	50.0	1.23174	1.229567	27.28
41.2	1.18472	1.182625	22.61	50.2	1.23284	1.230668	27.39
41.4	1.18575	1.183660	22.72	50.4	1.23395	1.231770	27.49
41.6	1.18679	1.184696	22.82	50.6	1.23506	1.232874	27.60
41.8	1.18783	1.185734	22.93	50.8	1.23616	1.233979	27.70
42.0	1.18887	1.186773	23.04	51.0	1.23727	1.235085	27.81
42.2	1.18992	1.187814	23.14	51.2	1.23838	1.236194	27.91
42.4	1.19096	1.188856	23.25	51.4	1.23949	1.237303	28.02
42.6	1.19201	1.189901	23.36	51.6	1.24060	1.238414	28.12
42.8	1.19305	1.190946	23.46	51.8	1.24172	1.239527	28.23
43.0	1.19410	1.191993	23.57	52.0	1.24284	1.240641	28.33
43.2	1.19515	1.193041	23.68	52.2	1.24395	1.241757	28.44
43.4	1.19620	1.194090	23.78	52.4	1.24507	1.242873	28.54
43.6	1.19726	1.195141	23.89	52.6	1.24619	1.243992	28.65
43.8	1.19831	1.196193	24.00	52.8	1.24731	1.245113	28.75
44.0	1.19936	1.197247	24.10	53.0	1.24844	1.246234	28.86
44.2	1.20042	1.198303	24.21	53.2	1.24956	1.247358	28.96
44.4	1.20148	1.199360	24.32	53.4	1.25069	1.248482	29.06
44.6	1.20254	1.200420	24.42	53.6	1.25182	1.249609	29.17
44.8	1.20360	1.201480	24.53	53.8	1.25295	1.250737	29.27

(continued)

Specific Gravity, Degrees Brix, and Degrees Baumé of Sugar Solutions, continued

Degrees Brix or % by Weight of Sucrose	Specific Gravity at 20/20°	Specific Gravity at 20/4°	Degrees Baumé (Modulus 145)	Degrees Brix or % by Weight of Sucrose	Specific Gravity at 20/20°	Specific Gravity at 20/4°	Degrees Baumé (Modulus 145)
54.0	1.25408	1.251866	29.38	63.0	1.30657	1.304267	34.02
54.2	1.25521	1.252997	29.48	63.2	1.30778	1.305467	34.12
54.4	1.25635	1.254129	29.59	63.4	1.30898	1.306669	34.23
54.6	1.25748	1.255264	29.69	63.6	1.31019	1.307872	34.33
54.8	1.25862	1.256400	29.80	63.8	1.31139	1.309077	34.43
55.0	1.25976	1.257535	29.90	64.0	1.31260	1.310282	34.53
55.2	1.26090	1.258674	30.00	64.2	1.31381	1.311489	34.63
55.4	1.26204	1.259815	30.11	64.4	1.31502	1.312699	34.74
55.6	1.26319	1.260955	30.21	64.6	1.31623	1.313909	34.84
55.8	1.26433	1.262099	30.32	64.8	1.31745	1.315121	34.94
56.0	1.26548	1.263243	30.42	65.0	1.31866	1.316334	35.04
56.2	1.26663	1.264390	30.52	65.2	1.31988	1.317549	35.14
56.4	1.26778	1.265537	30.63	65.4	1.32110	1.318766	35.24
56.6	1.26893	1.266686	30.73	65.6	1.32232	1.319983	35.34
56.8	1.27008	1.267837	30.83	65.8	1.32354	1.321203	35.45
57.0	1.27123	1.268989	30.94	66.0	1.32476	1.322425	35.55
57.2	1.27239	1.270143	31.04	66.2	1.32599	1.323648	35.65
57.4	1.27355	1.271299	31.15	66.4	1.32722	1.324872	35.75
57.6	1.27471	1.272455	31.25	66.6	1.32844	1.326097	35.85
57.8	1.27587	1.273614	31.35	66.8	1.32967	1.327325	35.95
58.0	1.27703	1.274774	31.46	67.0	1.33090	1.328554	36.05
58.2	1.27819	1.275936	31.56	67.2	1.33214	1.329785	36.15
58.4	1.27936	1.277098	31.66	67.4	1.33337	1.331017	36.25
58.6	1.28052	1.278262	31.76	67.6	1.33460	1.332250	36.35
58.8	1.28169	1.279428	31.87	67.8	1.33584	1.333485	36.45
59.0	1.28286	1.280595	31.97	68.0	1.33708	1.334722	36.55
59.2	1.28404	1.281764	32.07	68.2	1.33832	1.335961	36.66
59.4	1.28520	1.282935	32.18	68.4	1.33957	1.337200	36.76
59.6	1.28638	1.284107	32.28	68.6	1.34081	1.338441	36.86
59.8	1.28755	1.285281	32.38	68.8	1.34205	1.339684	36.96
60.0	1.28873	1.286456	32.49	69.0	1.34330	1.340928	37.06
60.2	1.28991	1.287633	32.59	69.2	1.34455	1.342174	37.16
60.4	1.29109	1.288811	32.69	69.4	1.34580	1.343421	37.26
60.6	1.29227	1.289991	32.79	69.6	1.34705	1.344671	37.36
60.8	1.29346	1.291172	32.90	69.8	1.34830	1.345922	37.46
61.0	1.29464	1.292354	33.00	70.0	1.34956	1.347174	37.56
61.2	1.29583	1.293539	33.10	70.2	1.25081	1.348427	37.66
61.4	1.29701	1.294725	33.20	70.4	1.35207	1.349682	37.76
61.6	1.29820	1.295911	33.31	70.6	1.35333	1.350939	37.86
61.8	1.29940	1.297100	33.41	70.8	1.35459	1.352197	37.96
62.0	1.30059	1.298291	33.51	71.0	1.35585	1.353456	38.06
62.2	1.30178	1.299483	33.61	71.2	1.35711	1.354717	38.16
62.4	1.30298	1.300677	33.72	71.4	1.35838	1.355980	38.26
62.6	1.30418	1.301871	33.82	71.6	1.35964	1.357245	38.35
62.8	1.30537	1.303068	33.92	71.8	1.36091	1.358511	38.45

(continued)

Specific Gravity, Degrees Brix, and Degrees Baumé of Sugar Solutions, continued

Degrees Brix or % by Weight of Sucrose	Specific Gravity at 20/20°	Specific Gravity at 20/4°	Degrees Baumé (Modulus 145)	Degrees Brix or % by Weight of Sucrose	Specific Gravity at 20/20°	Specific Gravity at 20/4°	Degrees Baumé (Modulus 145)
72.0	1.36218	1.359778	38.55	81.0	1.42088	1.418374	42.95
72.2	1.36346	1.361047	38.65	81.2	1.42222	1.419711	43.05
72.4	1.36473	1.362317	38.75	81.4	1.42356	1.421049	43.14
72.6	1.36600	1.363590	38.85	81.6	1.42490	1.422390	43.24
72.8	1.36728	1.364864	38.95	81.8	1.42625	1.423730	43.33
73.0	1.36856	1.366139	39.05	82.0	1.42759	1.425072	43.43
73.2	1.36983	1.367415	39.15	82.2	1.42894	1.426416	43.53
73.4	1.37111	1.368693	39.25	82.4	1.43029	1.427761	43.62
73.6	1.37240	1.369973	39.35	82.6	1.43164	1.429109	43.72
73.8	1.37368	1.371254	39.44	82.8	1.43298	1.430457	43.81
74.0	1.37496	1.372536	39.54	83.0	1.43434	1.431807	43.91
74.2	1.37625	1.373820	39.64	83.2	1.43569	1.433158	44.00
74.4	1.37754	1.375105	39.74	83.4	1.43705	1.434511	44.10
74.6	1.37883	1.376392	39.84	83.6	1.43841	1.435866	44.19
74.8	1.38012	1.377680	39.94	83.8	1.43976	1.437222	44.29
75.0	1.38141	1.378971	40.03	84.0	1.44112	1.438579	44.38
75.2	1.38270	1.380262	40.13	84.2	1.44249	1.439938	44.48
75.4	1.38400	1.381555	40.23	84.4	1.44385	1.441299	44.57
75.6	1.38530	1.382851	40.33	84.6	1.44521	1.442661	44.67
75.8	1.38660	1.384148	40.43	84.8	1.44658	1.444024	44.76
76.0	1.38790	1.385446	40.53	85.0	1.44794	1.445388	44.86
76.2	1.38920	1.386745	40.62	85.2	1.44931	1.446754	44.95
76.4	1.39050	1.388045	40.72	85.4	1.45068	1.448121	45.05
76.6	1.39180	1.389347	40.82	85.6	1.45205	1.449491	45.14
76.8	1.39311	1.390651	40.92	85.8	1.45343	1.450860	45.24
77.0	1.39442	1.391956	41.01	86.0	1.45480	1.452232	45.33
77.2	1.39573	1.393263	41.11	86.2	1.45618	1.453605	45.42
77.4	1.39704	1.394571	41.21	86.4	1.45755	1.454980	45.52
77.6	1.39835	1.395881	41.31	86.6	1.45893	1.456357	45.61
77.8	1.39966	1.397192	41.40	86.8	1.46031	1.457735	45.71
78.0	1.40098	1.398505	41.50	87.0	1.46170	1.459114	45.80
78.2	1.40230	1.399819	41.60	87.2	1.46308	1.460495	45.89
78.4	1.40361	1.401134	41.70	87.4	1.46446	1.461877	45.99
78.6	1.40493	1.402452	41.79	87.6	1.46585	1.463260	46.08
78.8	1.40625	1.403771	41.89	87.8	1.46724	1.464645	46.17
79.0	1.40758	1.405091	41.99	88.0	1.46862	1.466032	46.27
79.2	1.40890	1.406412	42.08	88.2	1.47002	1.467420	46.36
79.4	1.41023	1.407735	42.18	88.4	1.47141	1.468810	46.45
79.6	1.41155	1.409061	42.28	88.6	1.47280	1.470200	46.55
79.8	1.41288	1.410387	42.37	88.8	1.47420	1.471592	46.64
80.0	1.41421	1.411715	42.47	89.0	1.47559	1.472986	46.73
80.2	1.41554	1.413044	42.57	89.2	1.47699	1.474381	46.83
80.4	1.41688	1.414374	42.66	89.4	1.47839	1.475779	46.92
80.6	1.41821	1.415706	42.76	89.6	1.47979	1.477176	47.01
80.8	1.41955	1.417039	42.85	89.8	1.48119	1.478575	47.11

(continued)

Specific Gravity, Degrees Brix, and Degrees Baumé of Sugar Solutions, continued

Degrees Brix or % by Weight of Sucrose	Specific Gravity at 20/20°	Specific Gravity at 20/4°	Degrees Baumé (Modulus 145)	Degrees Brix or % by Weight of Sucrose	Specific Gravity at 20/20°	Specific Gravity at 20/4°	Degrees Baumé (Modulus 145)
90.0	1.48259	1.479976	47.20	95.0	1.51814	1.515455	49.49
90.2	1.48400	1.481378	47.29	95.2	1.51958	1.516893	49.58
90.4	1.48540	1.482782	47.38	95.4	1.52102	1.518332	49.67
90.6	1.48681	1.484187	47.48	95.6	1.52246	1.519771	49.76
90.8	1.48822	1.485593	47.57	95.8	1.52390	1.521212	49.85
91.0	1.48963	1.487002	47.66	96.0	1.52535	1.522656	49.94
91.2	1.49104	1.488411	47.75	96.2	1.52680	1.524100	50.03
91.4	1.49246	1.489823	47.84	96.4	1.52824	1.525546	50.12
91.6	1.49387	1.491234	47.94	96.6	1.52969	1.526993	50.21
91.8	1.49529	1.492647	48.03	96.8	1.53114	1.528441	50.30
92.0	1.49671	1.494063	48.12	97.0	1.53260	1.529891	50.39
92.2	1.49812	1.495479	48.21	97.2	1.53405	1.531342	50.48
92.4	1.49954	1.496897	48.30	97.4	1.53551	1.532794	50.57
92.6	1.50097	1.498316	48.40	97.6	1.53696	1.534248	50.66
92.8	1.50239	1.499736	48.49	97.8	1.53842	1.535704	50.75
93.0	1.50381	1.501158	48.58	98.0	1.53988	1.537161	50.84
93.2	1.50524	1.502582	48.67	98.2	1.54134	1.538618	50.93
93.4	1.50667	1.504006	48.76	98.4	1.54280	1.540076	51.02
93.6	1.50810	1.505432	48.85	98.6	1.54426	1.541536	51.10
93.8	1.50952	1.506859	48.94	98.8	1.54573	1.542998	51.19
94.0	1.51096	1.508289	49.03	99.0	1.54719	1.544462	51.28
94.2	1.51239	1.509720	49.12	99.2	1.54866	1.545926	51.37
94.4	1.51382	1.511151	49.22	99.4	1.55013	1.547392	51.46
94.6	1.51526	1.512585	49.31	99.6	1.55160	1.548861	51.55
94.8	1.51670	1.514019	49.40	99.8	1.55307	1.550329	51.64
				100.0	1.55454	1.551800	51.73

Reference

United States Bureau of Standards. 1942. Circular C 440. pp. 614, 626. Also published as AACC Approved Method 80-75.

Glossary

Aldehyde—An organic compound containing a -CHO group.

Aldose—A sugar molecule containing an aldehyde group at the terminal carbon position.

Baker's special—A granulation of sucrose with uniform crystal size and the ability to dissolve rapidly.

Bilateral symmetry—The balanced combination of left and right forms.

Caster sugar—An ultrafine granulated sugar useful for fine-textured cakes and meringues and, because it dissolves easily, for sweetening fruits and iced drinks.

Coarse granulated sugar—Sugar crystals slightly smaller than those of confectioner's sugar, often used for colorless or white candies.

Compound coatings—Coatings containing fats other than cocoa butter but similar to regular chocolate in melting properties.

Conching—Slow mixing of a heated chocolate paste to reduce particle size and increase thickness and smoothness.

Confectioner's sugar—A very large crystalline sugar, used primarily for its purity and very bright white color. It is a very hard crystal.

Creaming—High-speed mixing. In baking, the creaming step mixes fat and sugar.

Degrees Baumé (° Bé)—An arbitrary scale of specific gravities of liquids or solutions.

Degrees Brix (° Brix)—A measure of the density or concentration of a sugar solution. The degrees Brix equal the weight percent of sucrose in the solution.

Dextrorotary—Describing a compound that can cause a plane of polarized light to rotate in a clockwise fashion (to the right). Compounds that cause polarized light to rotate in a counterclockwise direction (to the left) are termed "levorotary."

Dextrose equivalent (DE)—A measure of the percentage of glucosidic bonds hydrolyzed. Dextrose has a DE of 100.

Dipole (dipolar)—A group of atoms having equal electric charges of opposite sign that are separated by a finite distance.

Disaccharide—A carbohydrate containing two sugar units, each composed of five or six carbon atoms in a furanose or pyranose ring.

Electrophilic—Having an affinity for negative charges.

Enrobing—Covering a base food material with a melted coating that hardens to form a solid surrounding layer.

Fondant—A grained confection often used as an ingredient in the manufacture of other candies such as fudge.

Fructose—A six-carbon keto sugar naturally present in fruits and honey and produced by the isomerization of glucose.

Furanose—A five-carbon sugar molecule in the form of a ring.

Galactose—A six-carbon aldo sugar and isomer of glucose; one of the two sugar units in lactose.

Gel grain—A granulation of sucrose coarser than powdered sugar yet finer than standard sugar. Used in dry mix applications.

Glucose—A six-carbon aldo sugar derived from starch and essential to human life.

Glycemic index—A measure of how much a food increases the glucose level in blood after it is digested.

Graining—Forming a crystalline structure.

Humectancy—The property of retaining moisture.

Humectant—A material that is able to attract water to itself.

Hydrophobic bond—An interaction, i.e., attraction, between two apolar groups in a polar (aqueous) environment.

Hygroscopicity—The ability to attract and retain moisture.

Invert sugar—A sugar consisting of equal parts of fructose and glucose that is made by the enzymatic breakdown of sucrose.

Isomerization—Conversion of a molecule from one isomeric form to another. An **isomer** is a molecule having components (number of carbons, hydrogens, etc.) identical to those of another molecule but with a different structural makeup.

Ketone—An organic compound containing a -CO- group.

Ketose—A sugar molecule containing the ketone group at the carbon molecule adjacent to the terminal carbon.

Lactose—A 12-carbon disaccharide, composed of one molecule of glucose and one of galactose, derived from milk.

Maillard browning—The browning of foods that occurs over time with high temperature. Also called **nonenzymatic browning** to distinguish it from the browning caused by an enzymatic reaction.

Maltose—A 12-carbon disaccharide, composed of two molecules of glucose, derived from starch.

Mannose—A six-carbon aldo sugar and isomer of glucose.

Massecuite—A dense mixture of sugar crystals and syrup that is an intermediate product in the manufacture of sugar.

Molasses—A thick, brown, uncrystallized syrup produced during the refining of sucrose.

Monosaccharide—A carbohydrate containing a single sugar unit, usually composed of five or six carbon atoms, existing in a furanose (five-membered ring) or pyranose (six-membered ring) form.

Nucleophilic—Having an affinity for positive charges.

Oligosaccharide—A carbohydrate containing four to seven sugar units (from the Greek *oligos*, meaning "a few").

Polysaccharide—A carbohydrate containing several hundred, thousand, or hundred thousand sugar units (from the Greek *poly*, meaning "many").

Powdered sugar—The finest granulation of sucrose crystals. Often used dry as a topping on doughnuts and candies or in fondants.

Pyranose—A six-carbon sugar molecule in the form of a ring.

Raw sugar—Sugar that has not undergone the refining process.

Reducing sugar—A sugar molecule in which the carbonyl group can react to form a carboxylic acid group. The sugar can undergo nonenzymatic browning.

Refractive index—A physical property of a substance that relates to how light is refracted from the material. Usually used to indirectly measure some other property, such as soluble solids (i.e., the total sugars in solution).

Sanding sugars—Sugars used as topical, dusted applications on bakery items and on gummed or gelled candies.

Short—Describing the texture of a product that breaks apart very easily when bitten.

Specific gravity—Ratio of the density of a sample solution and the density of water at the same temperature.

Stereochemistry—The relationship of atoms in three-dimensional space.

Sucrose—A 12-carbon disaccharide, composed of one molecule of glucose and one of fructose, obtained from sugar cane or sugar beets; the primary sweetener in the world.

Sugar alcohol or polyol—A compound derived by the reduction of sugar (in either the aldo or keto form), e.g., D-glucose to D-sorbitol.

Sugar bloom—A dusty white appearance on the surface of chocolate caused by the formation of certain types of sugar crystals.

Syneresis—The separation of a liquid from a gel; weeping.

Tastant—A substance capable of eliciting taste.

Trisaccharide—A carbohydrate containing three sugar units.

Turbinado sugar—A raw sugar that has been partially processed, removing some of the surface molasses. It is blond, has a mild flavor, and is often used in tea.

Water activity—The measurement of the degree to which water is bound in a system. The water activity of foods is measured on a scale of 0 (dry) to 1.0 (moist).

Xylose—A five-carbon aldo sugar derived from xylan hemicellulose, a by-product of paper pulp manufacture.

Zwitterionic—Having both positive and negative charges.

Index